Prove It

Elizabeth Finkel

Prove It

A SCIENTIFIC
GUIDE FOR THE
POST-TRUTH ERA

LA TROBE
UNIVERSITY PRESS
IN CONJUNCTION WITH BLACK INC.

Published by La Trobe University Press in conjunction with Black Inc.
Wurundjeri Country
22–24 Northumberland Street
Collingwood VIC 3066, Australia
enquiries@blackincbooks.com
www.blackincbooks.com
www.latrobeuniversitypress.com.au

9781760642716 (paperback)
9781743824207 (ebook)

A catalogue record for this book is available from the National Library of Australia

Cover design by Tristan Main
Cover image by Stocksy
Text design and typesetting by Aira Pimping
Index by Garry Cousins

To Alex, Archer and Victor, my compass points,
and Alan, the wind in my sails.

Dedicated to the memory of Lisa Lethin, Bella Hirshorn
and Harry Kestin, whose imprints are all over this book.

Contents

Preface

Some time ago, I realised my way of thinking was different from that of many people. I also realised my way of thinking was not "natural". It had been formed through my life in science, first as a working scientist, then as a science writer, and then as an editor of a popular science magazine.

This book is my attempt to share how science pursues knowledge, a mission embedded in the very word. Scientia is Latin for knowledge.

It's been an epic journey. My training was in the biological sciences, but as an editor, I had to be across everything. I began to intuit that there were differences in the way researchers operate. For instance, as a biologist I found what theoretical physicists do almost unrecognisable. And I was intrigued by the way changing technologies were changing the way science was done.

To explore these issues, I have journeyed across the sciences to show how five theories are being tested.

To make the cut, each of the five theories had to be impactful, relevant to the current moment, and importantly, it had to offer a good story.

Please note the title, *Prove It*, is meant in the sense of testing theories to destruction. Proof is not part of the vocabulary of science. As Michael Worobey, an evolutionary biologist at the University of Arizona, put it: "proof is for math".

Introduction

Scientia potentia est – knowledge is power.

Francis Bacon

How do we generate knowledge?

The question goes back to Socrates. He left no written works, but we hear his voice through the "dialogues" of his student Plato. Ever so gently, Socrates probes the beliefs of his conversation partner, methodically dissecting them to lay bare their logical fallacies. It has come to be known as the Socratic method.

One of Plato's most famous dialogues employs the allegory of a cave to teach Socrates' primary lesson: knowledge can be based on false beliefs. The cave is home to a group of prisoners who have been chained to a fence for their entire lives. All they have ever been allowed to see is the cave wall directly in front of them. Shadows of animals, plants and objects dance across the cave wall, representing the reality of the external world. The prisoners have no idea that the images are created by puppets paraded past a blazing fire just behind them. One prisoner breaks free and climbs out of the cave. Dazed by the sunlight, it takes time for his sensitive eyes to adapt. At first, he is only able to look at shadows, then reflections, then real objects. He dashes back to the cave to enlighten his fellow captives. But his eyes have not readjusted to the dark and he stumbles around. The prisoners perceive a blinded, deranged man, raving about a parallel world. They want nothing to do with him and become aggressive.

This is Plato's second lesson: the danger of trying to enlighten those wedded to pre-existing beliefs. Poignantly, Socrates would pay with his life for trying to enlighten others.

In a later dialogue, Socrates quizzes a young man by the name of Theaetatus. He probes the confident youngster's ideas on the nature of knowledge while gradually teasing out the knots in his logic. Continuing the lesson of the cave, Socrates starts by gently disabusing Theaetetus of his belief that the senses are a reliable source of knowledge. Next, he dismantles the young man's notion that arriving at a correct answer signifies knowledge: without an understanding of how one arrived at the knowledge, it is worthless. When Theaetetus suggests that beliefs accompanied by an understanding of the process is the route to knowledge, Socrates points out all the potential snags involved in truly understanding the process. The discussion is left for another day.

That day never comes, because Socrates heads off to his trial, where he is condemned to death for his "corruption of the Athenian youth" and "impiety".

Answering Socrates

It would take over 2000 years to come up with a satisfactory response to Socrates' questions. It arrived in the form of the scientific revolution. Its answer? Knowledge is best arrived at, not via the opinions of those with authority or influence, or even of those who make the most compelling arguments, but by putting theories to the test – the principle of empiricism. Those theories whose predictions are borne out by the results should be given the greatest weight.

Plato's student Aristotle also extolled the merits of evidence – "Credit must be given … to theories only if what they affirm agrees with the observed facts" – but Aristotle was more of an observer than an experimenter.[1] He was happy to entertain grand "untested" theories, such as the idea that the earth was the centre of the universe. By contrast, the scientific revolution was ushered in by the exacting astronomical measurements of Copernicus, Galileo and Kepler, which revealed that Earth and the other planets were in orbit around the sun, rather than the other way round. And let's not forget philosopher Francis Bacon, dubbed the "father of empiricism", for his 1620 magnum opus Novum Organon. It is a handbook on how to do science by carrying out exhaustive observations and drawing conclusions from those observations – a process called induction. His lively preface seems to leap across the centuries:

> Now my plan is as easy to describe as it is difficult to effect. For it is to establish degrees of certainty, take care of the sense by a kind of reduction, but to reject for the most part the work of the mind that follows upon sense; in fact I mean to open up and lay down a new and certain pathway from the perceptions of the senses themselves to the mind.

Bacon was also a savvy politician, a legal reformer and a master of maxims, like this one: "Money is a good servant, but a bad master."

Brilliant as all these astronomical and philosophical actors were, they were just the warm-up acts. The star role in the scientific revolution goes to Isaac Newton, who honoured his debt to those who came before with the timeless words: "If I have seen further, it is by standing on the shoulders of giants."

Standing on the shoulders of Copernicus, Galileo and Kepler , Newton glimpsed the sun-centred universe and pondered a new question: Why did the planets orbit the sun?

The French philosopher Descartes gave his answer in a 1633 book titled *The World*. He deemed that something like a giant tornado of dust particles raged around the sun, dragging the planets along with them. Isaac Newton was seven years old when Descartes died. By the time Newton was twenty-six, he was already the Lucasian Professor of Mathematics at Cambridge, no doubt for the stunning discoveries he made during the plague years, which he spent in isolation at his mother's farm in Lincolnshire, some 60 miles north of Cambridge. "Truth is the offspring of silence and unbroken meditation", he noted. His unbroken meditation gave birth to calculus (the maths of change), studies of optics (in the course of which he stuck a blunt needle into his eye-socket), his laws of motion, and the beginnings of his theory of gravity. Seeing an apple fall from a tree in front of the manor house was his Eureka moment. The force that made the apple fall to the earth, he mused, was the same as the one binding the planets to the elliptical solar orbits described by Kepler. Newton spelt out his laws of motion and gravity in 1687 in his most famous work, *The Mathematical Principles of Natural Philosophy*, better known by its Latin name, *The Principia*.

Today, most people have no problem with the idea of gravity as a force that pulls the apple to the ground or the earth to the sun. It was a different story in Newton's time. Descartes' tornado seemed the more rational view. How could the sun reach out across the vastness of space with some invisible force to pull on our planet? This was "barbaric physics" and could only be accounted for by a desire to be "returned to darkness", opined German mathematician and philosopher Gottfried Leibniz. Admittedly, he was peeved with Newton; Leibniz and Newton had rival claims as the first to develop calculus. Then, as now, Newton's was given primacy. But Leibniz was far from being the only one to label Newton's theory unscientific.

What vindicated the theory was that it made testable, precise predictions. It specified that the gravitational force between two objects increases with their masses and decreases as they grow further apart by the square of the distance between them. For example, if you double the distance, the force reduces to a quarter.

Newton's maths proved correct. It accurately predicted how long it would take for the moon to orbit the earth and the coming of Halley's comet. Newton's formula also predicted that the warped orbit of Uranus was due to the gravitational pull of a ghost planet. A century and a half later, Neptune was found. For 300 years, Newton's predictions kept hitting the mark. And for most earth-bound situations they still do.

Newton represents a watershed in the development of science. The peculiar thing about Newton, and what made him the lead actor of the scientific revolution, was that his theory, unlike those of Aristotle or Descartes, was limited to what could be accounted for by his mathematical predictions. "We are to admit no more causes of natural things than such as are both true and sufficient to explain their appearances," he wrote. Newton did not attempt to go beyond the data to explain what gravity is, or whether it really was a force at all: "I have not as yet been able to deduce from phenomena the reason for these properties of gravity, and I do not feign hypotheses."

This notion of science as being light on theory is familiar to me. As a working scientist (before I was a science writer, I was a molecular biologist), my contribution to theory was limited to what could be induced from my last successful experiment. I studied a gene that instructed the fruit-fly embryo how to construct its fourteen segments. I found that a part of the gene detected where it was in three-dimensional space, enabling it to lay down a series of stripes on the rice-grain-sized embryo, which at that stage was nothing but a sack of fluid. Those stripes provided the pattern for an industrious crew of cells to sculpt the segments.[2]

In my ten years as a working scientist, I never encountered the philosophy of science. Nor did I much encounter the topic in my decades writing about the work of other scientists. But in my journey for this book, which would see me roam widely, from theoretical physics to human evolution, and deeply, across the centuries, I knew I would have to reckon with the philosophy of science. I did not relish the task: reading philosophy can be challenging.

Moreover, I was not convinced that beyond Newton's strict empiricism, there was much philosophy at work in modern science. Indeed, according to Michael Strevens, a philosopher of science based at New York University, when scientists themselves are placed under the microscope to dissect their philosophical impulses, nothing coherent emerges beyond the compulsion to test, test, test. Strevens refers to this as "the iron rule".[3] As physicist Richard Feynman put it, "the philosophy of science is about as useful to science as ornithology is to birds". [4]

To my surprise, delight and relief, philosophy emerged unbidden from the very first chapter in the form of the Scottish enlightenment philosopher David Hume. It was fortuitous. Hume's philosophy provided a natural kick-off point for the chapters that followed.

What follows here is an introduction to the philosophical actors you'll encounter. It is a journalistic rather than academic treatment – a nod to the patron saints of philosophy who made cameo appearances on my journey through the sciences. As it happens, they are also the ones you'd likely meet in an introductory course of twenty-first-century philosophy of science, before the field becomes rather more crowded.

Philosophy of science

It was Albert Einstein who introduced me to David Hume.

Like other eighteenth-century Enlightenment philosophers, Hume valued individual reasoning over dogma and drew inspiration from the scientific revolution. In particular, he was inspired by Isaac Newton – "The greatest and rarest genius that ever wrote for the ornament and instruction of the species."[5] Newton inspired Hume. Hume in turn inspired Einstein to do what Newton could not: develop a theory of gravity.

It all began with the problem of relativity: namely that a light beam was not subject to the law of relative motion. For instance, as I stand here at my desk, my speed relative to the earth is zero, but my speed relative to the sun is 107,000 kilometres per hour. Every moving thing in the universe has a speed relative to something. But a light beam does not. In 1865, the Scottish mathematician James Clerk Maxwell showed that the speed of light was a constant 300,000 meters per second, regardless of anything else. How could that possibly be, Einstein wondered? In one of his famous thought experiments, he imagined peddling a bicycle to catch up to a light beam. Maxwell's equations said that no matter how fast he pedalled, the light beam would race away at exactly 300,000 meters per second.

Einstein discovered Hume in 1902 while working as a patent clerk in Bern, Switzerland. For fun, he and two colleagues, a philosopher and a mathematician, formed the Olympia Academy to discuss philosophy. They paid particular attention to Hume's 1739 *A Treatise of Human Nature*. Hume warned about the dangers of scientific induction, the practice of extrapolating from observations to formulate general laws of the universe. It may have been the method Newton employed, but it was an "intellectual habit", absent a solid philosophical foundation. As Hume wrote: "there can be no demonstrative arguments to prove, that those instances, of which we have had no experience, resemble those, of which we have had experience".[6]

A favourite example concerns the colour of swans. Since Roman times, the whiteness of swans was held to be a self-evident truth. But in 1697, Dutch sea captain Willem de Vlamingh, on a mission to look for shipwreck survivors on Australia's west coast, sailed up a river mouth and, lo, beheld black swans! The incident provided Perth's Swan River with its name, as well as a salutary philosophical lesson.

For Einstein, the lesson bore on the conundrum of the fixed speed of light. The metre and the second had always been considered unchanging quantities. But what if they weren't? What if space and time altered when a racing bicycle approached the speed of light? What if a metre shrank and a second grew longer? That would result in diminishing returns, because you'd be covering a shorter distance and taking longer to do it. Since speed equals distance over time, the peddler can never gain on the light beam. As described in the final chapter of this book, Einstein's ability to let go of his "intellectual habits" freed him to discover his theory of Special Relativity, and then General Relativity – his theory of gravity. Had he not read Hume, "I cannot say that the solution would have come", he wrote.[7]

Einstein freed himself from the intellectual habit of induction by using a "deductive" process instead. It relied not on observations but on the mathematical certainty of the constant speed of light. All very well for Einstein – but the vast majority of scientists do not have the luxury of starting from mathematical certainties. Most scientists operate by making observations about the real world and inducing theories from those observations. Other than 'Proving Relativity', every other chapter in this book describes how scientists deal with the problem of induction. And problems they are. While Einstein's theory of relativity has endured unchanged for more than a century, the same cannot be said of any of the other theories explored here. As one philosopher quipped: "Induction is the glory of science and the scandal of philosophy."[8]

Tackling induction

So how has science addressed the scandal of philosophy?

Hume himself suggested the solution: "Probability is founded on the presumption of a resemblance betwixt those objects, of which we have had experience, and those, of which we have had none."[9]

Probability and statistics, applied to lots of robust data, are the major remedy to the problem of induction. But there is also another, offered by Karl Popper.

I needed Einstein to introduce me to David Hume, but Karl Popper needed no introduction. He is the most famous philosopher of science of the twentieth century. If you've come across the idea that scientific theories can't be proven, only disproven or "falsified", that's courtesy of Popper.

Popper has a poignant personal story that resonates strongly with my motive for writing this scientific guide for the post-truth era. Born in 1902 into a cultivated, scholarly home – his mother a pianist, his father a lawyer – Popper's first decade was lived in Vienna's golden age. As the capital of the Austro-Hungarian Empire, Vienna was the seat of political power, but also a cauldron of European cultural and intellectual ferment. Here modernism exploded with the stylised eroticism of Gustav Klimt's shimmering gold paintings and the raw sexual canvases of Egon Schiele; the absurdist literature of Franz Kafka and the meltingly poetic works of Rainer Maria Rilke; the hauntingly beautiful music of Gustav Mahler and the atonal works of Arnold Schoenberg; the philosophy of Ludwig Wittgenstein; and of course Sigmund Freud's revolutionary theories about the life of the unconscious mind. "In those first fourteen years of the twentieth century, Vienna, more than anywhere else, was the fulminating, bewitching crucible where the modern world was invented," notes writer William Boyd.[10]

Popper witnessed its destruction. He was twelve when the First World War broke out and thirty-seven when the second one came around. In between, he flirted with and rejected Marxism, tried his hand at carpentry and teaching and managed to complete a PhD in the philosophy of psychology: "On Questions of Method in the Psychology of Thinking". With the rise of Nazism, his Jewish ancestry erased his job prospects. To build a reputation, he wrote a book, *The Logic of Scientific Discovery*. Published in 1934, it introduced his theory that the way to distinguish science from non-science is falsification. His ideas struck a chord and won him an offer to teach philosophy at Canterbury University College in Christchurch, New Zealand. He emigrated with his wife in 1937, a year before Austria was annexed by Hitler. In 1946, he moved to the United Kingdom to found the department of philosophy at the London School of Economics.

Popper experienced firsthand what can happen to the most intellectually progressive of civilisations when a populist ideology takes hold. How could a philosopher protect future generations from such an assault on truth? Like the enlightenment thinkers before him, his answer was the scientific method. "Truth is therefore the aim of science; science is the search for truth," he wrote.[11]

I was delighted to discover that Popper's theory – science searches for truth by falsifying theories – was inspired by Einstein! Popper heard Einstein expound on his astonishing theory of General Relativity in Vienna in 1919. Gravity was not a force but a consequence of the way mass causes a curvature in spacetime. A fantastical theory! But in the same breath, Einstein proposed a way to prove it wrong. During an eclipse, the moon blocks the sun, and the dark sky makes the stars near the sun suddenly visible. Although the stars themselves are very far away from the sun, their light rays have to pass close by it to be seen by people watching the eclipse. Einstein's prediction was that, like a banking snowboarder, the starlight would curve along the spacetime warped by the sun's huge mass. As a result, the apparent positions of the stars would be shifted by an exact amount predicted by Einstein's equations.

Bottom line: Einstein's theory could be falsified. As Popper put it, "Thus I arrived, by the end of 1919, at the conclusion that the scientific attitude was the critical attitude, which did not look for verifications but for crucial tests; tests which could refute the theory tested, though they could never establish it."[12]

Years later, Einstein's talk sparked Popper's key contribution to the philosophy of science. Science cannot prove theories, because, as Hume pointed out, what's true today may not be true tomorrow. Just because we observe a phenomenon once doesn't mean we can assume it will happen again. But science can certainly disprove things! That's what distinguishes scientific theories from, say, Freud's theory of the unconscious or Marx's theory of historical materialism – the idea that history is driven by economic forces and class struggle. Those theories do not offer falsifiable predictions. You might agree with one of these theories or disagree with it, but there is no way to disprove it. Science, by contrast, offers predictions that can be tested and therefore falsified. "I believe I have solved the problem of induction," Popper declared.[13]

Scientists prefer being right

Popper had his detractors. One was his former student Imre Lipschitz, a Hungarian-born philosopher who changed his name to Imre Lakatos, migrated to the United Kingdom after the Soviet invasion of Hungary in 1956, enrolled in his second PhD at Cambridge, became an assistant lecturer in Karl Popper's department at the London School of Economics, and by 1969 was its resident professor of logic. Notwithstanding his meteoric rise, Lakatos was never granted British citizenship, likely because of his former life as a Stalinist and informer to the Hungarian secret police.

Lakatos, like Popper, embraced the importance of falsification but argued that in reality, theories are rarely overturned by contradictory data. Rather, theories tend to accommodate the contradictions by virtue of a "protective belt" of auxiliary hypotheses. As he put it:

> Popper's criterion ignores the remarkable tenacity of scientific theories. Scientists have thick skins. They do not abandon a theory because facts contradict it. They normally either invent some rescue hypothesis to explain what they then call a mere anomaly and if they cannot explain the anomaly, they ignore it.[14]

The philosopher most diametrically opposed to Popper, and almost as famous, was the American Thomas Kuhn. No doubt you've heard the term "paradigm shift"? That's thanks to Kuhn. The term went viral thanks to his 1962 book *The Structure of Scientific Revolutions*, which sold over a million copies.

Kuhn earned a PhD in theoretical physics at Harvard University but was drawn to the history and philosophy of science, which he taught at Berkeley, Princeton and MIT. His seminal moment came when reading the scientific writings of Aristotle. He found Aristotle's ideas absurd, until he suddenly experienced his own paradigm shift and was able to view them through the prism of ancient times. For instance, Aristotle's understanding of motion included any type of change – not just the motion of the sun, but also the way it changed colour.

According to Kuhn, modern scientists were not so different. They too saw their scientific problems through the prism of a particular paradigm: "the entire constellation of beliefs, values and techniques" of their scientific community.[15] And rather than attempt to falsify their theories, as Popper proposed, they did the exact opposite: they designed experiments to affirm their working theory. When a theory is overturned, it is not through one fell swoop, but death by a thousand cuts of contradictory data. The overturning of a long-held theory in this way is what

Kuhn referred to as a "scientific revolution". But for the most part, "normal science" involves scientists working within their silos to shore up their dominant paradigm.

This blinkered determination to affirm existing theories accounts for much of the success of the scientific machine. For Kuhn, the reward for finding "proof" is held to be greater than for "disproof", which explains why scientists will spend their professional lives toiling over minutiae such as analysing the bumps on fossil human skulls, looking for more data to support their paradigm.

Patron saints

David Hume popped out of what was originally my first chapter, on Proving Relativity (it's now the last). But it turns out, with a little bit of nudging into the icon frame, each of my chapters can claim a "patron saint".

Karl Popper is the patron saint for the hunt for the origins of Covid-19. I was triggered to write this chapter because of some shocking journalism on the topic. If my book was meant to provide guidance for the post-truth era, there was no better example than this. It turned out to be a great decision, not least because some of the key scientists on the Covid origins case showed me that Popper is alive and well in the modern science lab. It is one of the very few times I have ever seen modern researchers guided by a philosophical compass. It may have had something to do with the fact that the eyes of the world were upon them.

"Popperian" scientists were amongst the first to propose that the virus came from a lab. Then they tried to see if they could disprove their theory. They succeeded. Much of the evidence they gathered failed to support the proposition that the virus was man-made; rather, it supported the theory that the virus originated in the wild-animal trade. Later, some of those same scientists tried to disprove that the Huanan seafood market in Wuhan, China, which also sold wild animals, was the source of the outbreak. Their attempts revealed the exact opposite. In the vast city of Wuhan, with its population of 12 million people and dozens of shopping malls and markets, the earliest infections clustered in only one place: around the Huanan seafood market.

Definitive proof for the origins of Covid may never be forthcoming. In the 2002 SARS outbreak – a kind of prequel to the Covid pandemic – the source was traced to infected palm civets sold at a Guangdong market. In the case of Covid, finding the animal source of the virus wasn't possible: shortly after the initial 2019 outbreak, the wild animals (most of them illegally traded) disappeared from the Huanan seafood market before they could be tested. Without that piece of the puzzle, sorting the origins of Covid comes down to a balance of probabilities. In January 2025, the Trump-appointed CIA director John Ratcliffe released an agency report claiming with "low confidence" that the virus most likely emerged from a lab. Ratcliffe said this new finding was not based on new evidence but on a fresh analysis.[16]

I can't speak to the CIA analysis. All I can do is follow the science. While there is plenty of speculation and joining of dots, as far as actual evidence for the lab theory, so far there is zero. By contrast, many different lines of evidence point to a natural origin for the virus. And here the philosophers of old, once again, help us out. The nineteenth-century British polymath William Whewell, who coined the term "scientist", also coined the term "consilience" to describe the principle "that evidence from independent, unrelated sources can 'converge' on strong conclusions". And while Popper did not believe *proof* was ever possible, he nevertheless championed the search for *truth*. His riposte to the philosophical relativists, who questioned the whole idea of truth, was to point out the difference between truth and certainty. Yes, truth might be elusive, but the real goal was to determine the degree of certainty. He likened the scientific process to the way jurors weigh evidence in a courtroom.[17] Try Covid's origins in a courtroom, and the evidence will show that there is little certainty about it being man-made. The balance of probability is on the side of a natural origin.

David Hume was patron saint for proving relativity. He is also patron saint for the chapter on human origins. But there's a difference. For relativity, Hume is framed as a cheerleader for catapulting Einstein out of his intellectual habits to discover the secrets of the universe. When it comes to human origins, Hume's portrait has a stern visage. This chapter is a cautionary tale of what befalls those who do not heed Hume's warning about the perils of induction. Charles Darwin provides the starting point for the chapter. He read Hume and was alive to the dangers of induction. "The crust of the earth with its embedded remains must not be looked at as a well-fitted museum, but as a poor collection made at hazard and at rare intervals," he wrote.[18] That's why he took twenty years to amass evidence from numerous species for his theory of evolution by natural selection. But when it came to the evolution of the human species, there was very little in the way of evidence – so-called "missing links" – on which to base his theory. Even as more human-like fossils were discovered, they nevertheless represented tiny pieces of an incomplete puzzle, and paleoanthropologists were inevitably tempted to guess at the whole picture. And so this chapter reads like a never-ending boxing match between competing theories. As one writer put it, the study of human origins is "the most emotional and trenchant of the sciences".[19]

But 150 years after Darwin, things have changed. Today, the big upheavals in understanding human evolution don't come from digging up fossils in caves but digging into human DNA. These new "computational geneticists" have a lot of data. They are talking the language of physics, notes Swedish geneticist Svante Pääbo, the Nobel Prize–winner who first revealed the DNA blueprint of a Neanderthal.[20] That's not to say the field has suddenly become all sweetness and light. There are new arguments over the algorithms used for analysing DNA and the startling things they suggest – among other things, that we carry DNA from a ghost lineage, perhaps a result of interbreeding with ancient *Homo erectus*.

Moving on from human origins, we come to the science of medicine – another field that has been plagued by the problems of induction. Its patron saint, Archie Cochrane, is not strictly a philosopher, although he was a great humanist.

For most of its history, medicine was not a science. Until the mid-nineteenth century, a visit from the doctor was more likely to kill than cure you. Bloodletting, hailed as a treatment for everything from fever to high blood pressure, persisted in the textbooks until the early years of the twentieth century. Science began its incursion into medicine, weakly, with experiments by Scottish ship surgeon James Lind, who tested treatments for scurvy in sailors in 1740. It would take another two centuries to weed out the devious statistical biases and human failings that consistently distort the results of such medical trials – an approach captured by a term coined in the 1990s: evidence-based medicine. Its central premise is that medical researchers should gather as many studies on a topic as possible and give more weight to those of higher quality.

Lumping different studies together is a valuable but perilous thing to do. The question of how to go about it goes back to seventeenth-century French mathematician Blaise Pascal, whose goal was to improve the odds for gamblers. But it is British doctor Archie Cochrane, the "father of evidence-based medicine", whom I've designated as this chapter's patron saint. He was a romantic figure. He volunteered to fight for democracy in the Spanish Civil War as a nurse and survived to finish his medical degree in London, only to enlist in the Second World War.

He was soon captured and spent four years in a German POW camp. As the sole doctor in the camp, he was horrified to find that his newly minted medical knowledge was just as likely to kill as to cure his fellow inmates. "I was afraid that I shortened the lives of some of my friends by unnecessary intervention," he later recalled.[21] It was a baptism of fire that forged his life's mission: to make medicine scientific. His story is vividly told in his memoir, *One Man's Medicine*. Today, the "Cochrane review", the highest standard of evidence for evaluating medical efficacy, is named in his honour. Its underpinning philosophy is that medicine should "do more good than harm".[22]

My chapter on the science of medicine shows that Cochrane's mission still has some way to go. It is a case study of the use of ivermectin, a worming medication, during the Covid-19 pandemic. Doctors around the world promoted its use despite the missing critical evidence: did it do more good than harm? It did not. Its widespread use in Peru likely contributed to Peru having the highest death rate from Covid in the world.[23]

Thomas Kuhn and Imre Lakatos are the patron saints of the chapter on testing consciousness. In opposition to Popper, both believed that scientists are primarily driven to prove rather than disprove their working theory. Both would find much to support their view in the newly fledged science of consciousness. Holed up in their siloes, some twenty camps vigorously defend their different theories. To advance the field, the Templeton Foundation, which is partial to big questions in philosophy and science, funded a series of "adversarial collaborations", in which competing camps collaborated to design experiments to put their opposing theories to the test. Ideally, these tests work according to Popper's dictum and falsify one or the other theory. In practice, they have generally played out in line with Imre Lakatos' observation: faced with contradictory data, the theories accommodate these results via auxiliary hypotheses. Either way, it's progress.

Many of the researchers studying consciousness are philosophers who have decided to add a scientific string to their bow. It goes with the turf: consciousness was philosophy's exclusive domain for 2000 years. I'd never before encountered such a marriage of science and philosophy. Modern-day scientists are often leery of the work of philosophers. But as we have seen in this introduction, philosophers – lovers of wisdom – have shepherded and sharpened the scientific endeavour since its inception. Socrates warned us of the limits of knowledge. Newton provided a template for relying only on testable evidence. Hume alerted us to the perils of induction – assuming that what is true today will always be true. Popper solved this problem by showing that science could proceed by disproving theories. Lakatos and Kuhn highlighted that science is a social activity, and one that rewards proof rather than disproof.

In the chapters that follow, you'll see traces of all these philosophical ideas at work.

1

Covid Origins

A scientist's job is to kick the tires of a hypothesis – to try to falsify it.

Michael Worobey[1]

According to twentieth-century philosopher Karl Popper, scientists aren't in the business of proving theories. Rather, they edge closer to truth by trying to disprove or *falsify* them. Those theories that withstand the scientists' battering ram are the best we have at any given moment. Of all the theories showcased in this book, those concerning the origins of Covid-19 provide the best example of Popperian science in action.

Since the early days of the outbreak, two theories vied to explain the origins of Covid. One is that like many outbreaks in recent decades, including HIV, Middle East Respiratory Syndrome (MERS), Ebola and the 2002 outbreak of SARS, it was a zoonosis: an animal virus that crossed into the human population. In this case, the crossing was linked to the Huanan seafood market in the city of Wuhan. Besides seafood, the market sold and butchered wildlife.

The other theory is that it leaked from a laboratory at the Wuhan Institute of Virology (WIV), most likely via an infected worker. The institute housed one of the world's largest collections of SARS-like viruses from bats and carried out experiments to see what they were capable of.

When it comes to the 2002 SARS outbreak, there appears to be no debate as to its source: palm civets at a Guangdong market. For the SARS-CoV-2 outbreak, however, the debate may never end.

Early evidence incriminated the market. Chinese authorities didn't wait for proof; they shut and sanitised the market on 1 January 2020. The wild animals, most of them traded illegally, vanished overnight. A month later, a federal law outlawed the farming of wild animals for meat, and the surrounding wildlife farms vanished too.[2] And so, the chance to find "animal zero", the species that may have harboured the virus, was lost.

As far as proving the lab origin, that evidence was also hard to come by. The Chinese, who were at first open to collaboration, steadily became less so.[3] A team of independent experts convened by the World Health Organization (WHO) visited Wuhan's markets and labs in January 2021, but negotiations for a second scheduled visit broke down.[4] Moreover, crucial DNA data revealing that raccoon dogs and civets, animals susceptible to SARS-CoV-2 , were traded at the market was withheld by the Chinese Centre for Disease Control and Prevention until March 2023.[5]

"These data could have – and should have – been shared three years ago," WHO director-general Tedros Adhanom Ghebreyesus fumed. "It's inexcusable. The scientific imperative, the public health importance, the moral importance of this should override everything else that's happening and it's not", added Maria Van Kerkhove, an epidemiologist who oversees the WHO program on emerging diseases.[6]

No doubt toxic politics helped poison the atmosphere for scientific collaboration. In March 2020, a tweet by President Donald Trump referred to "the China virus".[7] In May, his then trade adviser, Peter Navarro, accused China of seeding the pandemic by sending "hundreds of thousands of Chinese on aircraft to Milan, New York and around the world".[8] For some, China's stonewalling was tantamount to a confession of guilt. "I will just make the observation that the Chinese government, it seems to me, has been doing its best to try to thwart and obfuscate the work here," then FBI bureau director Chris Wray said in March 2023.[9] On the other hand, secrecy is the default position for Chinese authorities. It took four months for the world to learn about the 2002 SARS outbreak![10]

In the absence of proof, for the last four years the two theories have been subject to a battering ram whose ferocity would have pleased Popper. The battering has left the market theory standing and the lab theory in tatters. Let me take you through the gory details.

Beginning

With skyscrapers girdling the horizon, Wuhan in China's southeast is a megacity of some 12 million people. In December 2019, it served as the stage that made real the nightmare enacted in many a Hollywood movie: a pandemic that claimed nearly seven million lives and shut down the global economy for a year.[11]

It began with the sort of alert that makes public health officials jittery – "pneumonia of unexplained origin". Pneumonia is an infection of the lungs that causes inflammation and oozing. That oozing is a particular problem for the tissue of the lung, which is dotted with tiny air sacs less than half a millimetre in diameter known as alveoli. Pus clogs the sacs, preventing them from carrying out their major function: transferring oxygen to the fine mesh of blood capillaries that surround them. A patient can literally drown in their own body fluids.

In most cases, pneumonia originates from a resident bug, perhaps a pneumococcus bacterium or a cold or flu virus, that overruns the immune defences of a frail, elderly person. Doctors sometimes refer to it as old-man's friend. But if unexplained pneumonia takes hold in a young, healthy person, that's a red flag. It might signal the arrival of a deadly new strain of flu, like the one first reported in Mexico in 2009 that triggered the swine flu pandemic responsible for 240,000 deaths.[12] Or a new form of SARS – the severe acute respiratory syndrome that emerged in 2002 in Guangdong, China and claimed 774 victims.[13]

So, in December 2019, doctors in Wuhan were alarmed. At first, all they knew was that CT scans of the patients' lungs shared a distinctive feature known as "ground glass opacity". It referred to the hazy grey patches that appear when the lung's alveoli fill with fluid. Attempts to identify a new bug from the sputum of sick patients failed. Then, on 5 January Zhang Yongzhen – a virologist at the Chinese Center for Disease Control and Prevention and close colleague of virologist Eddie Holmes at the University of Sydney – identified the genetic code of a new virus. No doubt Zhang was following the advice of fifth-century BCE military strategist Sun Tzu: "Know thine enemy!"

Other Chinese scientists had also read the code of the new virus, but Zhang and Holmes were the first to release it to the world, on 10 January 2020.[14] Knowing the genetic code allowed the world to battle the virus by developing rapid tests to track infected people and deliver a vaccine in the record-breaking time of ten months.[15]

We had no such defences in 1981, when HIV emerged as a mysterious deadly disease disproportionately affecting gay men and stealthily contaminated the US blood supply. It took four years to read the genetic code of that virus.[16] It took five months to read the first SARS virus, and just days for the Wuhan virus.[17] To what do we owe this jump to warp speed? "Metagenomics": a technique that allows researchers to piece together the genetic code of a single species from fragments of DNA, even though the fragments are mixed up with the DNA of different species. It is the major tool of the trade for Zhang and Holmes.

Zhang and Holmes have collaborated to discover new viruses using metagenomics since 2011. To cut a long story short, they collect tissue samples and faeces from various creatures, extract the shredded genetic codes from thousands of viruses and then let computers piece the sequences back together. To get a sense of how fantastic a feat that is: imagine tipping the contents of 1000 boxed jigsaws of famous artworks into a heap, mixing them up – and then trying to piece together Renoir's *Luncheon of the Boating Party*, da Vinci's *Last Supper*, van Gogh's *The Starry Night* and so on …

Holmes and Zhang were focused on unearthing new animal viruses, but occasionally they were asked to identify new human viruses. On 3 January 2020, Zhang was asked to search for new viruses in the lungs of patients with pneumonia of unexplained origin. He and his team toiled for forty hours straight.[18] The novel genetic sequence they pieced together from one patient proclaimed the new virus to be a member of the SARS family, about 80 percent similar to the pandemic virus of 2002. On the morning of 5 January, Zhang contacted Holmes in Sydney. They agreed on two things: that the information should be immediately released to the Chinese health authorities, and that they would submit a paper to the prestigious journal *Nature*.[19] With lightning speed, they sent their draft on 7 January.

But there was a sticking point. The very day Zhang had received the sample, Chinese authorities issued an order forbidding labs from publishing information about the virus.[20]

If SARS left a lesson for the world, it was that every day counts in trying to contain a pandemic. Releasing the genetic sequence would allow sensitive PCR tests to enable countries everywhere to check individuals for Covid-19 infections and try to contain the disease's spread. By the morning of 11 January, Holmes could wait no longer. He called Zhang, who was boarding a plane from Shanghai to Beijing. Holmes insisted they release the full genetic sequence. Zhang asked for a minute to think about it, then called back and agreed to send Holmes the full sequence.

Within fifty-two minutes of receiving it, Holmes sent the sequence to Andrew Rambaut in Edinburgh, another of his comrades in the battle against viral outbreaks, particularly HIV and Ebola. Although it was close to 1 am in Edinburgh, Rambaut was up and eager to post the genetic sequence to his widely read virus bulletin board, Virological.org.

The day after, Zhang's laboratory at Fudan University in Shanghai was shut down by the authorities for "rectification", and he is still suffering the consequences.[21] "He has no real job and no real research lab. He has been effectively silenced as a scientist. Very sad," Holmes shared in an email in September 2024. Zhang's contribution was recognised by *Nature*, who named him one of the most influential scientists of 2020.[22] The release of the sequence not only equipped the world with PCR tests; it also launched the lightning-fast development of a new generation of RNA vaccines. The Moderna vaccine set a world record by commencing human trials sixty-nine days after the release of the genetic code.[23] All in all, a triumph for science.

The genetic code of the Wuhan virus revealed it to be a relative of the 2002 SARS virus. It was also a coronavirus, a name that reflects its crown-like "corona", which is visible under an electron microscope. And like SARS, rather than inscribing its code in DNA, it employed the dialect of RNA. But the family relationship was not close; the two viruses were only 80 percent similar. The WHO dubbed it SARS-CoV-2, short for SARS coronavirus 2.

Equipped with the code, labs in Wuhan deployed PCR tests to rapidly diagnose people infected with the new virus. And an alarming finding emerged. Most of the cases had a close connection to the Huanan Seafood Wholesale Market. Indeed, the first reported case on 10 December was a female seafood vendor.[24]

No surprises

For one international team, a new SARS outbreak was no surprise. For years they'd been warning that SARS-like viruses were poised for another jump into the human population. (As had China's top medical scientist, Zhong Nanshan.[25]) This team comprised two academics: Shi Zhengli at the WIV and Linfa Wang at the National University of Singapore. A third member, Peter Daszak, a British zoologist, headed a New York–based not-for-profit group dedicated to preventing pandemics. Known as the EcoHealth Alliance, it was founded in 1971 by famed British author and zookeeper Gerald Durrell. Now led by Daszak, its programs range from surveying wild animal populations to finding alternative jobs for wildlife smugglers.

In the wake of the 2002 SARS outbreak, Shi, Wang and Daszak joined forces to hunt down its source. SARS had been a wakeup call. While the world was alive to the pandemic risk of flu viruses, coronaviruses were considered agents of nothing more sinister than the common cold. SARS changed all that. One in ten patients died, a death rate 100 times higher than that of seasonal flu.[26] SARS spread from the Chinese province of Guangdong to a Hong Kong hotel and from there to twenty-nine countries.

The saving grace of SARS was that nearly all the infected became ill and took themselves out of circulation. That broke the chain of infection. Some seven months after it began, the WHO declared the pandemic over in July 2003. It was imperative to prevent such a virus ever emerging again.

So where had it emerged from? SARS first made itself known in February 2002, among restaurant workers in Foshan City in Guangdong province.[27] The restaurants served delicacies such as civet and snake soup, catering to *yewei*, a newfound taste for wild animal flesh. In May 2003, the restaurant supplies were traced to a wet market in the megacity of Shenzhen, some two hours' drive to the southeast of Foshan, across the Zhujiang River. As well as domesticated animals, it sold beavers, badgers, hares, fox-like raccoon dogs and weasel-like palm civets.[28]

Figure 1. Raccoon dogs (like this one) and civets can carry SARS-CoV-2. Their DNA was found at the Huanan seafood market in stalls that carried high levels of the virus.

Some of these animals came from wildlife farms, a government-supported industry that began in the 1950s with small-scale Sika deer farming, intended to reduce poverty while meeting traditional Chinese medicine's insatiable demand for antler velvet.[29] The ebbing of the Cultural Revolution saw growing entrepreneurship in the industry, catering not just to traditional Chinese medicine but also to a growing global market for fur and for animals for laboratories and zoos. The growing popularity of *yewei* saw the appearance of live wild animals at wet markets, ready to be butchered and barbecued at adjacent eateries.

By the 1980s, farmed animals included frogs, turtles, bamboo rats, civets, raccoon dogs, birds, snakes, pangolins, alligators and bears. There'd even been unsuccessful attempts to farm tigers and rhinoceroses. But Peter Li, an expert on China's animal policy based at the

University of Houston-Downtown, says many "farms" served as a laundering operations.[30] It was cheaper to trap animals from the wild. Transported in stuffy trucks, a Noah's Ark of creatures would be stacked in cages one above the other. Arriving at the markets, hundreds of species might be stored in poorly ventilated warehouses. The handlers took no precautions, some of them sporting open cuts. As Li told me, "every part of China's wildlife trade is a breeding ground for pandemic viruses".

At the Shenzhen market, researchers from the University of Hong Kong found that four palm civets and one raccoon dog were infected with a strain of the SARS virus similar to that seen in the earliest human cases. This strain carried twenty-nine letters of genetic code missing from later strains that circulated in human populations in Hong Kong, Guangdong, Vietnam and Canada. Like a tag of origin, the twenty-nine letters identified civets and raccoon dogs as the likely source of the SARS virus.[31]

Civets soon became the prime suspect. At one Guangzhou market, antibody tests showed that 80 percent of them had been infected. A thousand civets were culled and the wildlife trade was banned.[32] But not for long. Powerful business interests saw that ban overturned just three months later. By 2017, the wildlife trade was an eye-popping $77 billion dollar industry, greater than the GDP of Sri Lanka![33]

Soon, however, researchers realised that civets could not be the original source of SARS – the hotly sought Animal Zero. Civets sickened when infected by SARS. The species that provided the long-term reservoir for the new virus probably would not. That's what we see with bird flu. Wild water birds harbour the flu virus but often do not get sick; the virus then spreads to domestic birds, which come down with symptoms.

Linfa Wang, then based at Australia's CSIRO Animal Health Laboratory in Geelong, had a strong suspicion about what the reservoir species for SARS might be. In 1994, a mysterious flu-like illness had rapidly claimed the lives of eleven horses and their trainer in the Brisbane suburb of Hendra.[34] Wang was a member of the CSIRO team that identified the cause of the illness – a new virus distantly related to measles. It was named the Hendra virus.

After two years of dead ends testing the blood of possums, chickens, dogs, cats and other creatures for the Hendra virus, detective work finally led them to the source. It turned out that just a month before the main outbreak in Hendra, there had been a second, smaller outbreak 1000 kilometres to the north. Only a flying animal, and very likely a mammal, could have spread the virus that distance in such a short time.[35]

Fruit bats! The horses had probably grazed on bat-nibbled fruit.

Bats had not been on anybody's radar. It had been 100 years since epidemiologists last paid them any attention as a potential reservoir for a virus that infects humans.[36] That was the rabies virus.

Now, bat viruses were once again on the move. Four years after Hendra, in 1998 Nipah virus, which causes respiratory illness and has a fatality rate of up to 75 percent, crossed from bats to pigs and people in Malaysia. In 2005, Angola saw an outbreak of Marburg virus, a horrific "haemorrhagic fever" that causes bleeding from body orifices with a fatality rate of up to 90 percent. Its reservoir was traced to fruit bats in neighbouring Gabon. MERS, first reported in 2012, and with a mortality rate of 35 per cent, had crossed into humans from camels, probably decades ago, but its original source is believed to be bats. Ebola, a haemorrhagic virus related to Marburg virus with a mortality rate of up to 90 percent, caused outbreaks across West Africa from 2013–16 and is thought to have crossed over from bats before 1976.[37]

Why are bat viruses making these leaps into human populations now? The increasing connectivity of villages to large urban centres offers one explanation. But what is it about bats that make them reservoirs of such nightmarish diseases?

Wang became obsessed by the question. Bats are curious mammals. They're often resistant to cancer, and for a small mammal, they have an exceptionally long life span. While laboratory mice are lucky to make it to two years, the diminutive Brandt's bat, weighing in at less than a third their weight, lives for over forty years. But the truly distinguishing feature of this mammal is flight. Mammals like possums and squirrels take to the air by gliding, but only bats have evolved powered flight. Their marathon flights, up to 40 kilometres each night, put immense strain on their bodies. When marathon runners exert themselves to that degree, it takes them days or weeks to recover from the inflammatory effects. But bats rest from dawn to dusk, and then they repeat the performance. To keep going, Wang believes they have evolved a customised immune system – one that, unlike our own, is slow to inflame.

One of the side effects of a toned-down immune system might be that fruit bats react to viral infections differently.[38] When humans and horses were infected by Hendra virus, they died of pneumonia, an excessive inflammatory response. But bats seem to have found the sweet spot: they harbour the virus without excessive inflammation. Just how they do it is the subject of intense research by Wang and others. Bottom line: bats happily tolerate viruses that are fatal to other mammals.

So when SARS broke out in 2002, Wang immediately suspected bats. To track down the reservoir species, he joined forces with Peter Daszak and Shi Zhengli. After a decade of testing bat poo across Hong Kong, mainland China, Europe and Africa, in 2013 they found a needle in a haystack. A single bat cave was home to a colony of horseshoe bats that carried the ancestors of SARS.[39]

Shitou Cave lies some 60 kilometres outside the city of Kunming in Yunnan province. The southern edge of Yunnan province stretches across China's border with Myanmar, Laos and Vietnam, close to the infamous opium-producing region, the "Golden Triangle". This explains why Yunnan is famous not just for its limestone caves and stone forests, but also as the gateway for smuggling drugs and wildlife into China. Drug trafficking and prostitution also explain why Yunnan is the epicentre of HIV in China.[40]

The bat poo of Shitou Cave carried viruses that are 95 per cent similar to SARS, a closer match than that of any other bats sampled till then. But what clinched the family tie was a tiny snippet of code just twenty letters long known as the "receptor binding domain". Like the homing device on a missile, the receptor binding domain guides the virus to the surface of human lung cells – specifically to the ACE2 receptor. The ACE2 receptor is an important component of our body machinery, used to regulate blood pressure and inflammation. But the virus has commandeered it as its personal docking site.

The Shitou Cave viruses were tantalisingly close to explaining the 2002 outbreak of SARS. Less than 2 kilometres from a village, it was easy to imagine the chain of transmission. In the course of their nightly foraging, bats often roost on the thatched roofs of animal pens, where their virus-laden faeces drop down to infect the animals below. From there, the virus could make its way to the animal markets.

In 2013, the findings earned the team a paper in *Nature*.[41] But they were not done yet. The Shitou cave virus was only 95 per cent similar to the 2002 SARS virus. How had the virus acquired its final 5 per cent of code? Over the next five years, they sampled more bat faeces from Shitou Cave. In 2017, they got an answer that shocked them.[42]

Fifteen different strains of virus were shuffling chunks of genetic code between them. While no single strain was identical to SARS, all the building blocks of the 2002 SARS pandemic virus were accounted for. Most worryingly, that viral orgy was producing novel offspring. What might they be capable of?

To find out, Shi Zhengli in Wuhan now joined forces with virus genetic engineer Ralph Baric, based at the University of North Carolina. Their experiments showed that the new viruses were capable of infecting human cells. The risk of another outbreak loomed large.

Peter Daszak's EcoHealth Alliance distributed pamphlets to villagers, warning them of the danger and educating them about "Living safely with bats".[43] In scientific journals, the researchers published papers that warned of the clear and present danger. Like this one from 2015: "A SARS-like cluster of circulating bat coronaviruses shows potential for human emergence"[44]. And this from 2017: "Our current study therefore offers a clearer picture on the evolutionary origin of SARS-CoV and highlights the risk of future emergence of SARS-like diseases."[45]

Two years later, in December 2019, SARS CoV-2 struck the city of Wuhan. Just as prophesied.

Gain of function

For some it was the fulfilment of a quite different prophecy.

As it happens, the year 2017, when the source of SARS was finally traced to Shitou Cave, was also marked by an auspicious event 1000 miles to the northeast. After fourteen years of planning and construction, the WIV proudly launched its state-of-the-art, level 4 biosecurity labs, the sort where scientists in spacesuits do their work tethered to dayglo-blue spiral air hoses, and which can only be accessed through two separate airlocks with self-locking doors.

The US$44 million lab, situated well above the flood plain and able to withstand a magnitude 7 earthquake, would focus on the world's greatest biological threats, including African nightmares such as Ebola and the Lassa virus (China had about a million people working in Africa, and viruses do not respect borders), as well as homegrown horrors such as SARS. As part of a global network, it would act as a WHO "reference laboratory". "Our contribution on the BSL4-level pathogens will benefit the world," effused George Gao, director of the Chinese Center for Disease Control and Prevention. "The world is facing more new emerging viruses, and we need more contribution from China."[46] It was a lofty moment for international collaboration.

Not everyone shared that perspective.

Some scientists were nervous about a Chinese lab carrying out research on the world's greatest biological threats. For one thing, they pointed to the significant risk of leaks. A 2007 British outbreak of livestock foot-and-mouth disease was linked to effluent from a lab in Surrey, and SARS had already leaked twice from labs in China.[47] For another, they were alarmed that experimenting with viruses could create the very nightmare they were trying to prevent. Indeed, it had happened just six years before.

While most wild bird species can carry bird flu with impunity, it is deadly to other species. For poultry, the mortality rate approaches 100 percent; for humans, 60 percent.[48] The saving grace is that while an infected bird can infect a human, an infected human does not readily infect other humans. For this reason, most outbreaks have been limited to poultry handlers. Since 2003, there have been some 1100 human deaths globally.[49]

What keeps flu researchers up at night is that bird flu might acquire the ability to transmit between humans. With the aim of developing an early warning system, two researchers set out to discover what sort of code changes might give the virus that ability. In 2011, they found out. Yoshihiro Kawaoka, based in the USA, and Ron Fouchier, in the Netherlands, mutated portions of the H5N1 bird flu virus and infected ferrets, which, like humans, spread flu by sneezing. They used different methods, but their findings were the same: only a handful of mutations were needed for the virus to be spread by sneezing ferrets.

It was an electrifying moment. As Fouchier told *Science*, one of the world's highest-ranking journals, in 2011, the lab-made virus was "probably one of the most dangerous viruses you can make".[50]

And thus began a scientists' forever war. One camp argued that such "gain of function" experiments should never have been done. But given they had been, the results should never have been published, lest they become recipes for bioterrorists. Others argued that these experiments provided essential intelligence for the early warning system against pandemic flu.[51] Scientists who screen flu viruses each year would be alert to dangerous code changes, and preventive vaccines might be developed. Indeed, just as predicted, in March 2024 a new strain of bird flu began spreading among US dairy herds, and virologists avidly monitored it to see if it was acquiring any of the code changes highlighted by the ferret experiments.[52]

The counterargument: the potential benefit was not worth the risk of an accidental or terrorist-sparked pandemic. "The only impact of this work is the creation, in a lab, of a new, non-natural risk", argued Rutgers University–based molecular biologist Richard Ebright.[53] The US National Institutes of Health (NIH) placed a temporary ban on the experiments. Anthony Fauci, then head of the National Institute of Allergy and Infectious Diseases (NIAID), supported it.

In 2017, the NIH attempted to resolve the debate by introducing a safeguard. "Gain of function" experiments would be evaluated on a case-by-case basis, weighing the potential benefits against the risks before they could go ahead.[54] But the safeguard applied only to US government-funded research, not to privately funded research or to US-funded research in other countries. Some experts warned that risky research could still be carried out in countries with poor lab safety standards or a culture of secrecy. "Diversity of viewpoint, flat structures where everyone feels free to speak up and openness of information are important," noted one biosecurity consultant.[55]

The outbreak of SARS-CoV-2 in the city that housed the WIV was just the catastrophe some of these scientists had predicted.

Market or lab?

As SARS-CoV-2 began spreading in the city of Wuhan, contact tracers tried to track its source. Half of the earliest cases worked at the Huanan seafood market or had close contact with someone who did. On 1 January 2020, Chinese authorities shut and disinfected the market. On 27 January, it was reported that samples swabbed from the walls and cages of the market had tested positive for the virus. The vast majority (thirty-three out of thirty-five positive samples) came from the stalls where wild animals were sold.[56]

"The virus stems from wild animals on sale at the market," proclaimed Xinhua, the official Chinese news agency, in late January. On 24 February, Chinese authorities permanently banned the consumption and illegal trading of wildlife to "eradicate the bad habit of indiscriminate consumption of wildlife [and] effectively prevent major public health risks", reported Xinhua.[57] The wildlife farms were shut down, farmers compensated, and the animals disappeared – either freed or killed.[58]

It seemed China was quite clear about the origins of SARS-CoV-2: it had jumped ship from a wild animal at the Huanan seafood market. Yet by February 2020, the world was already entertaining another hypothesis: the lab leak.

A laboratory origin?

The rapid release of the genetic code of SARS-CoV-2 on 10 January 2020 was a triumphal moment for science. It also triggered a decidedly less-triumphal narrative.

As virus experts around the world interrogated the genetic code, they found it was about 80 percent similar to SARS. But it was the differences that compelled them. The new SARS appeared to be spreading more quickly than its older cousin. Could the code differences supply a clue as to why?

An obvious place to look was the region that coded for the "spike", one of the protruding components that form the virus crown. The spike is more armament than ornament. Like a homing device on a missile, it determines what type of cell the virus will lock onto. Zeroing in on the spike, they saw changes, precisely in the region that fine-tunes the homing device. These changes likely explained why the virus spread so avidly between people. They were also very precise. Could the virus have been tweaked in a lab?

A buzz began to grow louder on biologists' chat sites. The WIV was a world-renowned sentinel for monitoring bat viruses. Its freezers contained some 2000 such viruses (the vast majority of them as shredded fragments of RNA), and it conducted genetic engineering experiments to test their function. Could such a test have gone awry?

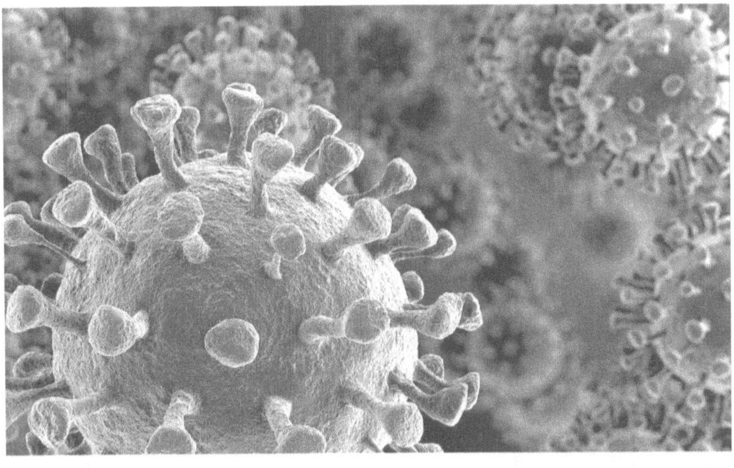

Figure 2. SARS-CoV-2 seen in all its menacing glory through an electron microscope. The surface spikes latch on to target cells. The RNA blueprint is safely tucked inside the outer membrane.

Jeremy Farrar, then head of the London-based Wellcome Trust, the world's largest funder of infectious disease research, tuned in to the buzz. On 28 January, he put that question to the person he considered the world authority on virus evolution: Eddie Holmes.

Holmes reasoned that if the new virus was created at the WIV, the institute must have housed a parent virus that was far more closely related to SARS-CoV-2 than SARS (which was only 80 percent similar). As it happens, Shi Zhengli, the leading bat virus researcher at the WIV, had just revealed that such a virus was housed in the institute's freezer.[59]

Frantic that her lab might be responsible for an accidental leak, she had screened all the genetic codes for the bat coronaviruses in her freezers. One gave her pause. Isolated in 2013, "RaTG13" was a 96.2 percent match to SARS-CoV-2. Its name revealed its provenance: "Ra" designates that it came from the horseshoe bat *Rhinolophus affinis*, "TG" that the bat was caught in the town of Tongguan in Yunnan province, and "13" that it was caught in the year 2013. Yet, despite its close relationship to SARS-CoV-2, Shi was confident that RaTG13 could not be the parent virus.

Most of the viruses in her freezer, like this one, had been retrieved from bat faeces as fragments of genetic code. To bring the virus to life required coaxing the virus to multiply in cells (typically African green monkey cells) in the petri dish. Only if it was brought to life and multiplied would it be possible for such a virus to escape from the lab, or to serve as the parent for genetic engineering experiments. Of the 2000 bat viruses retrieved by Shi's group, only three had ever been brought to life this way. RaTG13 was not one of them.[60]

Shi immediately released an online description of RaTG13 as a preprint.[61] Holmes burrowed into Shi's paper to answer Farrar's query. Could the new virus have been engineered using RaTG13 as the starting chassis?

He zeroed in on the crucial code for the receptor binding domain. It was nothing like the one carried by the new SARS virus. His response to Farrar was basically "nothing to see here".

Two days later, Holmes' equanimity was upended by a message from Kristian Andersen: "I've seen something very strange in this sequence, can you take a look?"[62] Based at Scripps Research in La Jolla, California, Andersen was another comrade-in-arms from campaigns to track down the origins of Ebola, Lassa, Zika and other viruses.

Andersen's message

Andersen considered it "possible and even plausible" that the virus had accidentally escaped from the lab at WIV. "We needed to consider it as a potential hypothesis, because of the work they were doing, and the proximity to the outbreak," he told me during a Zoom interview in April 2024.

Andersen was familiar with the work of the Wuhan lab because of its prolific publications in high-ranking journals such as *Nature*. It employed standard genetic engineering techniques to identify whether SARS-like bat viruses could infect human cells. If these techniques had been used to create the new virus, the hallmarks should be detectable. He zeroed in on the genetic code of the spike.

The strategy of any virus is to hack the operating system of the host cell so that instead of running its own programs, the host cell produces copies of the virus. For SARS-CoV-2, the first step requires docking with the host cell, which the virus achieves by virtue of the receptor binding domain (RBD) on the end of its spike. Then insinuating its viral malware – its genome – into the cell requires exposing a second component of the spike, one which fuses the viral envelope and cell membrane together as if they were two soap bubbles, allowing the viral genome to slither in.

And now we get to another storied aspect of the SARS-CoV-2 spike – the furin cleavage site (FCS). What is an FCS? In order for the spike to fuse with the cell membrane, its fusion machinery, which is normally hidden within the spike protein, has to be exposed. That is achieved by snipping the spike protein at a precise site, the FCS. Like dotted lines on a cardboard cut-out model, the FCS guides an enzyme, furin, to the cleavage site. Once cut, the fusion machinery flips up like a jack-in-the-box. Furin is provided by the host cell – yet another case of the virus helping itself to the host's machinery for its own nefarious ends. Sometimes the fusion machinery can be exposed without the help of an FCS; the original SARS did not have one. But the presence of the FCS on SARS-CoV-2 increased the infectiousness of this new virus. As we'll soon see, the FCS would take on huge significance.

When Andersen zoomed in for a closer look at the code of the new coronavirus, he focused on the spike. Two features stood out. Compared to the first SARS, the code for the RBD was altered. Second, unlike the first SARS, it carried an FCS. In addition to these suspicious differences, one more thing really got Andersen's pulse going. When genetic engineers stitch in a piece of new code, sometimes the "stitch", technically a "restriction site", remains visible. These stitches come in different types. One, known as "BamH1" was visible at one end of the spike code. Moreover, all the novel features of the virus lay on one side of this stitch; on the other it resembled its relatives. It seemed as if a new piece had been inserted and the stitch was a "junction point"

"Holy shit ... I came to the realisation that this could very well have been engineered", he tells me.[63]

From zero to 100 in five seconds

This was what was behind Andersen's "I've seen something very strange" message to Eddie Holmes. Infected by his colleague's alarm, Holmes told me, "We went from zero to 100 in the space of five seconds." Holmes circled back to Farrar in London, who contacted Anthony Fauci (then head of NIAID). Fauci recommended two things: that Andersen and Holmes "get a group of evolutionary biologists together ... very quickly", and that if this group validated their concerns, Farrar, Andersen and Holmes contact their national intelligence agencies. And so, without delay, on 1 February (or 2 February Sydney time), the hour-long conference call took place.

Besides Holmes and Andersen, there were Andrew Rambaut (the researcher who had posted the virus genome to his site Virological.org) and Robert Garry from Tulane University. The four men were close comrades from past campaigns where they worked together to track the source of deadly viral outbreaks such as Ebola. As science writer David Quammen put it, doing this kind of work, "sequencing strains ... of Ebola, as sampled from human patients *in extremis,* amidst a public health catastrophe, is a task for brainy people with big hearts and strong nerves. You make friends you can trust."[64]

The other participating experts included Marion Koopmans (known for tracing the origin of MERS to camels), Christian Drosten; Stefan Pöhlmann and Ron Fouchier (one of the scientists who showed how bird flu could become transmissible in laboratory ferrets). Also on the call were Anthony Fauci, Jeremy Farrar and Francis Collins, then head of the US NIH.

Like Holmes and Andersen, some of the participants came to the conference call freaked out by what Andersen had found. "Seeing that [furin cleavage site] in the new SARS-CoV-2 when those sequences came out for the first time actually kept me up all night," confessed Garry.[65] "It felt as if a storm was gathering, of forces beyond anything I had experienced and over which none of us had any control ... I was going to have to be persuaded that things were not as sinister as they seemed," wrote Jeremy Farrar.[66]

The call was a chance for the experts to challenge each other's hypotheses about the origins of the unusual features found in the spike. Andersen had emailed everyone a slide comparing the 30,000-letter codes of two viruses: the novel SARS virus above and RaTG13 below. He instructed his audience to focus on the region around the spike of the new virus, pointing out the changes in its RBD and FCS. He also pointed out the BamH1 stitch. Could the differences in the new virus have been man-made?

Maybe. But there was a more prosaic explanation. The BamH1 stitch is represented by six letters of genetic code. While the stitch per se was absent from the same region of code seen in the virus's closest relative, RaTG13, the difference was a single letter. Christian Drosten was "not convinced" that it was a stitch at all. In his view, it was far more likely that a single-letter mutation, the commonest kind, had accidentally created it. That was the inference the group of experts, particularly Fouchier, Koopmans and Drosten, leant toward.

This was only one of the arguments against the lab-made hypothesis. Another was that if the new virus had been genetically engineered by the Wuhan lab, it would have been generated using the standard techniques known to be employed there.

For one thing, if they had wanted to test the effects of an FCS, they could have started with one of the viruses that readily grew in their lab, such as WIV-1 or WIV-16, not some random bat virus that could not be grown. Likewise, if they had employed BamH1 stitching, why did the virus end up with this very odd-looking stitch?[67]

As Jeremy Farrar put it, "A malevolent scientist is still a scientist – and the most methodical way of conjuring up a nightmare virus would be to take a virus that is already a known quantity, such as SARS-CoV-1, and crank up its infectivity using known methods."[68]

Finally, what if the lab had found some way to grow RaTG13? Could it have served as the chassis to engineer this new virus? With its 96.2 percent match to SARS-CoV2, it might be tempting to think that a genetic engineer could simply have tweaked the RBD and inserted an FCS. But two arguments counter that temptation.

First, if a genetic engineer had knowingly designed the RBD, they would have done so using computer models based on the first SARS. But those models got it wrong; they predicted the RBD of the new coronavirus was "not ideal".[69] In reality, the RBD of the new coronavirus is, by some estimates, more than ten times better at doing its job than the SARS version.[70] The "unpredicted" success of the new virus RBD argued against human design. It was more likely to have emerged from the laboratory of nature – an argument that would soon be vindicated.

Second, even though RaTG13 was 96.2 percent similar to the new coronavirus, that left 1100 letters of genetic code separating the two, way too many for a genetic engineer to knowingly tweak.[71] Viruses are sleek machines that jettison bits of code that compromise their copy speed. Genetic engineers are lucky if they can coax a virus to take on a few new letters. Evolutionary biologists calculated it would have taken fifty years to evolve this degree of difference between the two viruses.[72]

More evidence

Following the 1 February conference call, the core group – Holmes, Rambaut, Garry and Andersen – agreed that genetic engineering was an unlikely explanation for the origins of the new virus. But there was analternative hypothesis for how the new virus could still have emerged from a lab. Not through genetic engineering, but by accelerated evolution in laboratory-grown cells. Andersen's suspicions had been aroused after a conversation with a colleague led him to believe that if coronaviruses were cultured in cells, they could evolve an FCS. If the new coronavirus had acquired the FCS this way, it would look indistinguishable from natural evolution. In other words, although made in a lab, it would not bear the hallmarks of genetic tinkering.

The next few days upended everyone's thinking. The first upheaval came two days after the conference call. Holmes received an email from his colleague Tommy Lam at the University of Hong Kong. A coronavirus from smuggled Malayan pangolins, seized in Southern China, had an RBD that was almost identical to the new one found in SARS-CoV-2. This vindicated the view that the RBD upgrade that gave SARS-CoV-2 a tenfold advantage over the original SARS was not the work of man but nature. "In the space of a couple of days, I changed my mind comprehensively," Holmes told me.

Andersen's holdout theory – that the virus might have evolved its FCS while multiplying in cells in the lab – took several more days to work through.[73] He trawled through the scientific literature, consulted colleagues and realised he was mistaken. There was no evidence that coronaviruses evolve an FCS when multiplying in cultured cells. In fact, some experts believed the opposite to be true: they lose the FCS when grown in cultured cells. While an FCS may be an advantage to a virus spreading in animals, it also makes the structure of the spike more flimsy. As Cornell University virologist Gary Whittaker put it, introducing the FCS into a virus in the lab generally causes it to "crash and burn".[74] Andersen thought a colleague had told him otherwise, but he now suspects he misunderstood. "Either I misheard it, or [my colleague] wasn't actually correct on that. That happens when you're still learning a lot about coronaviruses; you're going to get a lot of things wrong. This was one of them."

A third piece of evidence that swayed the group was a survey of the broader coronavirus family, the *Betacoronavirus* genus. It revealed that the FCS was present in many of them, including MERS and in those that cause the common cold, HKU.1 and OC-43.[75] What these viruses also have in common with SARS-CoV-2 is their propensity to jump species. If nature was evolving these FCS over and over again, especially in coronaviruses that crossed from animals to humans, why suspect a lab origin?

"It wasn't [just] one thing" that made the lab theory unlikely, Andersen told me. "The balance of probabilities was that the novel coronavirus had evolved naturally."

Two weeks later, on 17 February, Andersen, Holmes, Rambaut and Garry, plus a fifth comrade, Ian Lipkin – a veteran of Nipah and MERS outbreaks who provided scientific advice for the 2011 movie *Contagion* – published their analysis as a preprint. They concluded: "this analysis provides evidence that SARS-CoV-2 is not a laboratory construct nor a purposefully manipulated virus." But, they added, "it is currently impossible to prove or disprove the other theories of its origin described here, and it is unclear whether future data will help resolve this issue."[76] Their final paper was published in *Nature Medicine* on 17 March 2020. Titled "The Proximal Origin of SARS-CoV-2", it has been accessed more than six million times.

But not for the reasons the authors might have hoped. In good Popperian fashion, the authors had attempted to falsify their own theory of lab origins. While they had fallen short of proof, the evidence persuaded them against their initial suspicions. "Overall, this is a textbook example of the scientific method where a preliminary hypothesis is rejected in favor of a competing hypothesis after more data become available," Kristian Andersen told *The New York Times*.[77]

For others, rather than a textbook example of the scientific method, it was evidence of a conspiracy. The conspiracists noted that the authors had changed their view after the conference call on 1 February. They accused Fauci of persuading them to cover up the laboratory origins of Covid-19.[78]

What motive would Fauci have had? To avoid embarrassment because Fauci's institute had funded work at the WIV, said the conspiracists. And why would all the other people on the call – ten independent scientists from around the world – collude? American science journalist Nicholas Wade had an answer: they were engaging in an *omerta* to protect "gain of function" research.[79] Australian journalist Sharri Markson made similar accusations.[80] Moreover, Wade had uncovered another motive for Andersen to collude with Fauci. Andersen was awaiting the results of a US $9 million grant proposal submitted by himself and other researchers to the National Institute of Allergy and Infectious Diseases, which was headed by Fauci.[81]

Republicans in the US House of Representatives seized on the cover-up theory to stoke rage against Fauci in a series of subcommittee meetings that began in March 2023.[82] In one remarkable session, Andersen and Garry were confronted with emails and private Slack messages obtained from those who participated in the conference call. One phrase lifted from Andersen's email was touted as the smoking gun. It read: "disprove any type of lab leak theory". Clearly these Republicans had never heard of Karl Popper. As Andersen explained to the subcommittee, "I'm referring to the concept of what's called falsification."[83]

But it was not just journalists and politicians who stoked the conspiracy theory. Richard Ebright, the Rutgers-based opponent of gain-of-function research and his colleague Bryce Nickels, escalated this line of attack on social media, accusing Fauci, Andersen, Holmes and others who were part of the conference call, of being: "Fraudsters. Liars. Perjurers. Felons. Grifters. Stooges. Imbeciles. Murderers ..."[84]

Hardly the stuff of scientific debate. But for what it's worth, the conspiracy claims do not stack up.

Fauci says that at the time of the 1 February call, he had no idea that his national institute had provided funding to the WIV, pointing out that it funds projects to the tune of US$6 billion every year. The WIV received US$120,000 per year, subcontracted via the EcoHealth Alliance, an amount hardly likely to come to his notice.

Second, the claim that Andersen colluded with Fauci to protect his stake in a US$9 million grant, has a few inconvenient truths to contend with. NIAID grants are not awarded by fiat of Fauci. Grants are awarded through an independent, peer-reviewed process, and Andersen's grant, though not yet formally approved, had scored winning marks a month prior to the 1 February call.[85] Holmes, meanwhile, had missed out in a share of a similar grant – hardly consistent with the allegation that they were colluding for the grant money.

Disproving the market origins theory

As we have seen, the authors of "The Proximal Origin of SARS-CoV-2" – the most influential paper in the Covid origins debate and begetter of a never-ending conspiracy theory – concluded that on the balance of probabilities, SARS-CoV-2 was born of nature, not a lab. But it did not claim to have proven this absolutely or rule out the possibility of other origins. Indeed, some of the authors went on to try to disprove the theory that the virus crossed from animals into humans at the Huanan seafood market. To do this, they joined forces with Michael Worobey, an evolutionary biologist at the University of Arizona.

For Worobey, the falsification of theories is his ruling scientific creed. One colleague described him as having a "soft spot for wild theories".[86] No doubt he was referring to Worobey's extraordinary efforts to nail the origins of HIV. Two decades ago, most scientists believed HIV had crossed into humans from wild chimps, perhaps during the butchering of bush meat. A contrary theory held that the virus originated from a lab – specifically, one that was developing a polio vaccine. Chimpanzee cells had been used to grow the polio virus used to make the vaccine. In theory, those chimp cells might also have harboured the chimp ancestor of HIV. And so it was possible that HIV did not cross into humans from wild chimps, but through the point of a needle during polio vaccinations in northeastern Congo in the late 1950s.[87] Although stocks of the original polio vaccine were checked and did not carry HIV, some scientists remained unconvinced.

There was a way to falsify the vaccine origin of AIDS theory. The Democratic Republic of the Congo (DRC) was home to two subspecies of chimp: the western and the eastern. The western subspecies carried viruses closely related to HIV, supporting the theory the virus had jumped species. However, it was the eastern subspecies that has been used to manufacture the polio vaccine. So did the eastern subspecies carry HIV-like viruses? In early 2000, the second Congo war was raging. Who was going to risk a trip to the jungles of eastern DRC to collect chimp poo?

Michael Worobey, William Hamilton (an Oxford evolutionary biologist famous for the theory of the selfish gene) and evolutionary geneticist Jeff Joy at the University of British Columbia would. Together with armed guards, they trekked into the forests near Kisangani on the bend of the Congo River. They retrieved chimp poo but paid a high personal price. Worobey impaled his hand on a gigantic rattan spike, and as the wound started to rot, he nearly died. Hamilton did die. After contracting jungle malaria, he succumbed to complications shortly after making it back to the UK. Their tragic expedition helped lay to rest the polio vaccine theory of the origins of HIV.[88] Neither the eastern chimps nor the vaccine made from their cells were the source of HIV.

Worobey and four of the authors of "The Proximal Origin of SARS-CoV-2" would apply that same dedication to attempting to falsify the market origin theory of SARS-CoV-2.

Unsatisfied by the WHO report

In March 2021, the WHO released a report by a team that had visited Wuhan, touring the WIV labs and the Huanan seafood market to study the origins of SARS-CoV-2. Their report concluded that a market origin was "likely to very likely", while a laboratory origin was "extremely unlikely".[89]

That left a number of scientists unsatisfied. In a letter to *Science* the following May, they wrote that the "two theories were not given balanced consideration" and called for scientists to do more to explore the possibility of a laboratory origin.[90]

One of the signatories of the letter was Harvard-based molecular biologist Alina Chan, a long-standing proponent of the lab-leak theory. In a May 2020 preprint, she'd argued that SARS-CoV-2 could not have crossed over from an animal at the Huanan market because it was too well-adapted to humans. Five years later it remains a preprint; evidently her argument was too weak to pass peer review.[91]

Worobey was also one of the signatories to the letter to *Science* magazine; indeed, he'd initiated it. Now, he went looking for stronger evidence to falsify the market origins theory. During the Covid-19 lockdown, he secreted himself away in his basement and, with the help of Google Translate, spent three months searching Chinese newspaper reports, official documents and every scientific study that might shed light on the origins of the pandemic.

When it comes to tracing the source of a pandemic, the earliest patients are the most informative, because they can help scientists to pinpoint the source before the virus has diffused out into the world. Early reports linked twenty-seven of the first forty-one hospitalised cases to the Huanan seafood market. Most of these people worked at the market, either as vendors or delivering goods.[92] The very first known case was a woman who sold seafood who reported feeling unwell on 10 December.[93]

Yet these cases did not *prove* that the market was the source of the virus. Two types of counterarguments were mounted. The first concerned "ascertainment bias". If researchers have a preconceived idea of the sort of result they're searching for, they are more likely to find that very result. In the case of Covid, everyone knew that the 2002 SARS virus had jumped to humans in a wild animal market in Guangdong. So this time around, patients who said they worked at one of the city's wildlife markets might have been the first to be tested. This could have distorted the picture, making it seem that the earliest confirmed cases were market workers.

The second counterargument was that around a third of the early patients had no link to the market. That wasn't really surprising; many people were asymptomatic, so the virus could easily have spread through untraceable contacts, perhaps on a train or at a crowded mall. Nevertheless, it was possible the virus could have originated at one of the two campuses of the WIV which lie 12 and 27 kilometres to the south of the market, across the Yangtze River. Perhaps an infected scientist from one of the campuses had shopped at the market and spread the virus. In this scenario, the market was not the source of the virus but the site of a superspreader event.

Worobey set out to test these possibilities using information from the earliest Chinese hospital publications. He could find no evidence of ascertainment bias in the first forty-one cases.[94] The patients had been identified based on their clinical symptoms before the link with Huanan seafood market was identified on 29 December. But what about the cases who had no link to the market? The WHO report found that among the first 168 cases, only fifty-five were associated with the market, leaving two thirds with no link. Perhaps their home addresses would shed new light?

For 154 of the cases, Worobey was able to pin down the latitude and longitude of their home addresses. So did these point to one of the WIV labs? They didn't. Astonishingly, in a city of 12 million people, studded with dozens of densely packed shopping malls, each with 100 times more traffic than the Huanan seafood market, all the dwellings mapped in a ring around the market. The closer you got to the market, the higher the density of cases. "Long story short, they all lived absurdly close to, and centred on, the wildlife market," Worobey told me on a Zoom call in September 2023.[95] It was textbook epidemiology, akin to the radial distribution of cases identified by John Snow in 1854, when he traced the source of a London cholera outbreak to the Broad Street water pump.

No other location in Wuhan showed this bullseye pattern for early cases, certainly none of the campuses of the WIV. By February, just two months after the outbreak began, the pattern changed, with cases most likely to reside in the densely populated parts of the city, particularly those housing older residents.

Worobey's report was published in *Science* in July 2022. Its co-authors included Holmes, Garry, Rambaut and Andersen.[96] Worobey and his colleagues had set out to falsify the market origin theory. Instead, they ended up providing the strongest evidence to support it!

Animal ghosts

Researchers working on the 2002 SARS outbreak were able to close the case because they had the bodies – infected civets and raccoon dogs from the Guangdong market. In the case of SARS-CoV-2 the bodies were missing: the wild animals vanished when the market was shut down on 1 January 2020. And yet the Huanan seafood market, although abandoned and ghostly, continued to deliver evidence.

Scientists from the Chinese Center for Disease Control took swabs from the market in the days after it was closed. The genetic material in those swabs revealed three key pieces of information. One was that, contrary to previous claims from the Chinese CDC, wild animals, including raccoon dogs – suspected carriers of SARS-CoV-2 – were present in the market.[97] The other was that the swabs positive for SARS-CoV-2 clustered in the wildlife section, with a hotspot in one stall. Here a cage, a cart for moving animals, and a machine for removing hair and feathers, returned high readings of the virus.

"We are seeing the DNA and RNA ghosts of these animals in the environmental samples, and some are in stalls where SARS-CoV-2 was found, too," explained Florence Débarre, the French evolutionary biologist who accidentally stumbled on the CCDC market data in March 2023 and went on to analyse it after it was publicly released a few weeks later.[98] "The market covers a couple of acres, and this [cluster of positive swabs] comes down to one corner of the market, and to a couple of stalls," added Dominic Dwyer, a virologist at Sydney's Westmead Hospital who was part of the WHO team that visited Wuhan in 2021. "That fits with an animal origin. If it was coming from people wandering around the market, you'd find it everywhere."[99]

And it wasn't just one strain of virus that was detected in the market samples. In the early days of the pandemic in Wuhan, two main lineages of virus were isolated, A and B. Strangely, the first patients carried either one or the other, and didn't carry intermediates of the two. That was not consistent with one strain being the parent of the other. Rather it suggested they had two independent origins.[100] That interpretation was strengthened by finding that both strains were already present in the market before it was shut down. If, as some claimed (including the Chinese CDC), the market was merely the site for a superspreading event by an infected visitor, then the visitor would have contaminated the market with a single strain. The fact that both strains were present is just what you'd expect if the virus had jumped ship multiple times from different infected animals. This is what took place in the 2002 Guangdong SARS outbreak.

A final piece of information about the virus's ancestry was provided by the virus itself. Analysing the genetic codes of hundreds of the earliest isolates of SARS-CoV-2 has allowed researchers to arrange them on a family tree and trace its roots. Various analyses agree that the most recent common ancestor of SARS-CoV-2 crossed into the human population sometime between late November and early December 2019.[101] These time stamps match the first reported case on 10 December: the seafood vendor at the Huanan seafood market. This timing jives with tests of blood donated to the Wuhan Blood Center, which showed no antibodies to SARS-CoV-2 prior to December 2019.

All these independent lines of evidence point to the Huanan seafood market as the source of the outbreak. As Worobey put it, "No scenario other than sales at the wildlife market can explain all the data."

A bomb by the name of DEFUSE

By 2022, Worobey had painted a coherent picture of how SARS-CoV-2 emerged at the Huanan market, but a buried bombshell threatened to blow it up. Ironically, it was a bombshell by the name of DEFUSE.

It was first unearthed in September 2021. A group calling themselves DRASTIC (Decentralized Radical Autonomous Search Team Investigating Covid-19) used a freedom of information request to unearth a failed 2018 grant proposal to the US Defense Advanced Research Projects Agency (DARPA). DRASTIC claimed the proposal was a recipe for cooking up Covid![102]

The proposal, named DEFUSE, was an ambitious, $14 million, three-and-a-half-year project aimed at "defusing" the disease threat of bat-borne viruses. Its grand plan ran across seventy-five pages and had two arms. The first was to predict the disease risk of different bat viruses. This involved testing their spike genes, to see how well they enabled the virus to enter human cells in the lab. It also involved testing the blood of people who lived in proximity to bats for antibodies to the spikes carried by these viruses. "Our modelling team will use these data to build machine-learning genotype-phenotype models of viral evolution and spillover risk," the proposal said. That is: predict what different spike genes will do in the real world. The second arm aimed to develop a vaccine or otherwise boost the immune systems of bats so that they were less likely to harbour and shed dangerous viruses.

DEFUSE brought together five research leaders with different expertise from around the world. Three were based in the US: Ralph Baric at the University of North Carolina (genetic experiments and bat vaccines), Jerome Unidad at Palo Alto Research Center (immune boosters for bats), and Tonie Rocke at the US Geological Survey Wildlife Health Center (bat monitoring). Two were based outside the US: Linfa Wang at Duke–NUS (the immune system of bats) and Shi Zhengli at the WIV (collecting viruses and carrying out genetic experiments).

It was Ralph Baric's proposed experiments to test the function of spike genes that raised suspicions. They included tweaking the spike to see how that affected the virus's ability to spread in human airway epithelial cells. In one paragraph, there was a reference to testing the effects of potential furin cleavage sites.

> We will analyze all SARS-CoV S gene sequences … for the presence of potential furin cleavage sites … Where clear mismatches occur, we will introduce appropriate human specific cleavage sites and evaluate growth potential in Vero cells [monkey kidney cells] and HAE [human airway epithelial] cultures.[103]

This work was to be carried out in Baric's lab in North Carolina.

For some, this single paragraph was tantamount to a smoking gun. None of the immediate ancestors of SARS-CoV-2 had an FCS. And here was a proposal to insert one: "we will introduce appropriate human specific cleavage sites …"

The grant was rejected by DARPA, but that was no guarantee the work did not proceed. As Alina Chan commented in September 2021, "This was a highly funded group of researchers that wouldn't let one rejection halt their work."[104]

The intelligence community leans in

DEFUSE detonated at the centre of the Covid-19 origins debate. The fallout was massive. Amongst other things, it led to the US Congress passing the *COVID-19 Origin Act* in March 2023.[105] It required the Director of National Intelligence (DNI) to "declassify any and all information" relating to links between the WIV and the origins of Covid-19 within ninety days of the law's enactment.

In dispassionate, measured language, the June 2023 report from the DNI summarised the findings of nine different intelligence agencies.[106] "All agencies continue to assess that both a natural and laboratory-associated origin remain plausible hypotheses to explain the first human infection," the report stated.

But there were differences of opinion. The FBI and the Department of Energy leaned towards a "laboratory-associated incident". The National Intelligence Council and four others leaned towards natural origins. The CIA and an unnamed agency were undecided. The DNI offered an explanation: "Variations in Intelligence Community analytic views on the origins of the Covid-19 pandemic largely stem from differences in how agencies weigh intelligence reporting and scientific publications and intelligence and scientific gaps."

For all the "variations", the summary made the following three points:

- Almost all IC [intelligence community] agencies assess that SARS-CoV-2 was not genetically engineered.
- All IC agencies assess that SARS-CoV-2 was not developed as a biological weapon.
- ... some of the research conducted by the PLA [People's Liberation Army] and WIV included work with several viruses, including coronaviruses, but no known viruses that could plausibly be a progenitor of SARS-CoV-2.

Crucially, it is this last statement that most weighs against the hypothesis raised by the DEFUSE grant: namely, that a virus held at the WIV was the progenitor of SARS-CoV-2.

In July 2023, Eddie Holmes inadvertently provided some of that intelligence. Five years earlier, Holmes and Jie Cui, a Chinese researcher who had previously worked in Holmes' lab at the University of Sydney, had surveyed bat coronaviruses held at WIV, amongst them RaTG13. Holmes, Cui and colleagues at the WIV (including Shi Zhengli) had tried to publish the results in a journal but were rejected because the genetic codes of the viruses were incomplete. Instead, they contributed the codes to the GenBank genetic database and forgot about them. In July 2023, when the database publicly released those codes, Australian journalist Sharri Markson seized on their existence as more evidence of a "cover-up", penning an article titled: "Edward Holmes claims bad memory for not declaring writing a paper with a Wuhan scientist".[107] But what the forgotten paper showed was that the WIV held no viruses that could have served as the precursor of SARS-CoV2. "The irony is, I'd loved to have remembered this," Holmes told me.

Most virologists agree with the US intelligence community assessment that no virus at WIV could have been the precursor for SARS-CoV-2.[108] RaTG13, the closest known relative at the time of the outbreak, still differed from SARS-CoV-2 by 1100 letters of code. And switching that amount of code is simply beyond the capabilities of a genetic engineer. Stuart Turville at the University of New South Wales engineers viruses to act as delivery vehicles for therapeutic genes. "The odds of being able to artificially introduce such huge chunks of code are extremely low," he told me.[109]

The new chunks of SARS-CoV-2 are all-important to the success of the virus, but even now, virologists have little understanding of their role. For Cornell University virologist Gary Whittaker, it points to this: "SARS-CoV-2 was the design of nature, not man." Or as Linfa Wang put it, "no scientist, smart or rich, would be able to make SARS-CoV-2" until nature had shown us how it was done.[110]

Even if you believe, as the lab-leak champions do, that Shi Zhengli at the WIV is covering something up – she has, for instance, declined to surrender her WIV lab books – there can be no such suspicions about Ralph Baric. The intelligence agencies would have had free access to his North Carolina lab. Their finding was that no precursor virus existed.

Baric himself has been viciously attacked, accused either of making SARS-CoV-2 and sending it to Wuhan, or enabling WIV researchers to produce the virus. The fact that he pioneered a technique that leaves no trace of genetic stitches – so called "no see" genetic engineering – is further fuel for his attackers.[111] But Baric, with all his state-of-the-art capabilities, agrees with the findings of the intelligence communities. When he was hauled in front of the US House subcommittee on 22 January 2024, he declared in sworn testimony: "There's no genome close enough that is engineerable using current standards that could have resulted in SARS2."[112]

The final evidence comes from the virus itself. It has had five years to roam freely across one of the most populous and interconnected species on the planet. A dream run for any virus. And what do we see? The fast-evolving virus is changing at the rate of thirty letters per year.[113] To "evolve" 1100 changes in the space of a year or so in a lab, as one theory would have required, seems way beyond what even this rapidly mutating virus is capable of. As Turville told me, "It's beyond the virus machinery to change that much over that time."

Searching for ancestors

It took ten years to track the closest relatives of the first SARS virus to Shitou cave in Yunnan province. For SARS-CoV-2, it took less than two. In October 2021, a French team found its kinfolk in a bat cave just south of the Yunnan border, in Northern Laos.[114]

Having sourced these viruses from the bat's anal passage, the French team dubbed them "BANAL". They were anything but. They scored a striking 97 percent code match to SARS-CoV-2. By contrast, the closest SARS relative found to date, also from a cave in Yunnan and named YN2020F, delivered a 96 percent match.

Another striking feature of these viruses was their predilection for human cells. Their homing device, the ACE receptor, outperformed SARS-CoV-2 in its ability to zero in on human cells. It was a striking disproof of Alina Chan's argument that SARS-CoV-2 was too well tuned to humans to be "natural".

But for all its striking family resemblance, BANAL was missing a key ancestral trait: the FCS. Its absence from the closest family members remains a missing piece of the puzzle of the origins of SARS-CoV-2.

In February 2023, Zhengli Shi's lab did identify a SARS relative that carried an FCS.[115] It was a MERS-like coronavirus isolated from pangolins. Too distant a relative to qualify as an ancestor of SARS-CoV-2, says Holmes. Nevertheless, the discovery confirms that in pangolins, coronaviruses gain an evolutionary advantage from carrying an FCS.

So, could pangolins have been the direct source of SARS-CoV-2? Probably not. For one thing, they were not present at the Huanan seafood market. But then how to reconcile the fact that pangolins keep turning up with SARS-like viruses? Holmes believes it's the result of "sampling bias". Pangolin scales are a prized but unproven ingredient in traditional Chinese medicine. Supplying the market means these critically endangered creatures are the most trafficked mammal in the world. "Sampling bias" comes in because they are also the species most often seized in anti-smuggling operations. Rather than being the species in which SARS-CoV-2 evolved its FCS, they are more likely the canary in the coal mine, an indicator of the dangers lurking in their environment. They are likely only one of many species in which SARS-like coronaviruses learn to adapt to different mammals.

Best theory

If not pangolins, then what? The horseshoe bats of a Laos cave carry the closest virus relatives so far, but they do not carry an FCS.[116] So where did SARS-CoV-2 complete its evolutionary journey?

In late February 2024, I asked Holmes for his best theory of the origins of SARS-CoV-2. His answer: the wildlife trade.

Holmes joins the dots as follows. The hunt for bat ancestors has revealed a biogeographic divide. SARS-1 relatives tend to be found in Yunnan province and to the north. SARS-CoV-2 relatives are found in Yunnan and across the southern border into Laos – a region that also corresponds to the habitat of the Malayan pangolin. An infamous illegal trade route runs from Myanmar east to Laos, north into Yunnan and from there to major cities such as Wuhan.

As for which species provided the finishing touches to the evolution of SARS-CoV-2 and passed it on to humans, it could have been any of those trafficked in the wildlife trade: raccoon dogs, civets, marmots, bamboo rats, Amur hedgehogs, hog badgers, to name a few – all of which also happened to be present in the southwestern corner of the Huanan seafood market immediately prior to the 2019 outbreak.[117]

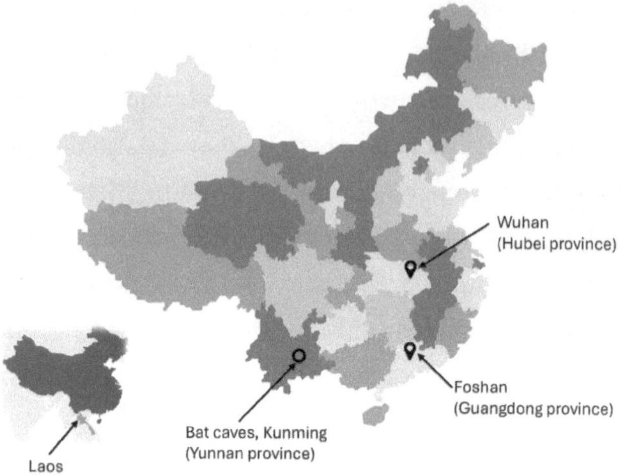

Figure 3. From bat caves to big cities. The 2002 SARS outbreak in Foshan was linked to bat viruses in Yunnan Province. The 2019 outbreak in Wuhan has been linked to bats in Yunnan Province and Laos. For both pandemics, the virus travelled over 1300 kilometres. Wild animals, smuggled from regions near China's southern border to markets in its major cities, are likely carriers.

Packed into crates, one on top of the other, transported long distances in trucks, a respiratory virus could jump from species to species, and some of its hopeful offspring might just gain a foothold; one might even have adapted to this ecosystem. It is an ideal cauldron in which to evolve a "generalist virus", which is one of the defining features of SARS-CoV-2. It has infected more than fifty animal species, from white-tailed deer herds in the forests of the USA to tigers and hippopotamuses in zoos.[118]

So, if the virus evolved in the wildlife caravan as it moved through Southeast Asia and into Yunnan, surely it would have infected local people. Indeed, villagers in Yunnan province carry antibodies that recognise SARS-like viruses, but it's impossible to say exactly which virus they were exposed to.[119] As to when the virus first emerged in Wuhan, tests of blood bank samples show that prior to December 2019, none carried SARS-CoV-2 antibodies. Nor do WHO flu alerts show any spike in flu cases prior to then. This is consistent with the timing estimated by tracing the virus family tree – placing its emergence sometime in mid-November.

Will we get any more evidence? I asked Holmes.

The obvious place to look is the wildlife trade. But it's illegal and secretive. Traffickers are not likely to open up their trucks to allow scientists in to test their animals. And at the official level, the Chinese government has shut down further inquiry. Peter Li, the animal policy expert at the University of Houston, believes that is partly a defensive posture against the toxic politics, and partly the influence of the trade itself. Once a US$77 billion industry, the wildlife trade is now fighting for survival. Li says the lab-leak theory has helped the trade to stage a "spirited comeback" because it exonerates the industry from blame.[120]

As Holmes told me in February 2024, as far as finding the origins of SARS-CoV-2: "We might get incremental evidence, but that's pretty much it. I think we're at the end of the road."

Postscript

On 3 June 2024, the day Anthony Fauci was yet again interrogated by a US House subcommittee, I was astonished to see a *New York Times* opinion piece with the snappy title: "Why the Pandemic Probably Started in a Lab, in 5 Key Points".[121] Highly polished and accompanied by stunning graphics, the authoritative-sounding article led one to believe that powerful new evidence had resolved the debate. What had I missed?

Nothing.

The article was by Alina Chan, and her five points had all been tested and refuted. *The New York Times* did not publish a direct riposte, but the letters summed it up. "Has science become a matter of opinion?" wrote Doug Allchin, a philosopher of science based in Minnesota. More detailed replies appeared in other publications. In *The Nation*, epidemiologist Gregg Gonsalves and immunologist John P. Moore described Chan's piece as "a rehashing of old and frequently rebutted arguments" and pointed out that Chan "has never worked on infectious diseases or the science of how viruses enter the human population, but she has become the go-to expert for proponents of the lab leak theory". *The Journal of Virology* ran an article by forty-one virologists titled "The Harms of Promoting the Lab Leak Hypothesis ... without Evidence". *The Lancet Microbe* ran an editorial titled "COVID-19 Origins: Plain Speaking Is Overdue and declared: "Being wrong, and fearlessly so, is part and parcel of robust enquiry, but those who platform these flawed arguments without responsible levels of scrutiny have serious questions to answer."[122]

Responding to this barrage of criticism, *New York Times* senior writer David Leonhardt defended Chan's piece by invoking Occam's Razor, which he defined as "a philosophical principle holding that the simplest explanation for a phenomenon is usually the correct one. In this case, a new SARS-like virus started in a city with one of the world's leading labs for researching SARS-like viruses."[123] Leonhardt might as well deploy Occam's Razor to argue that the moon landing didn't happen. What distinguishes science from conspiracy is that science rejects assertions that fail to clear the bar of evidence.

2

Human Origins

The Central Enigma

It has often and confidently been asserted, that man's origin can never be known. But . . . it is those who know little, and not those who know much, who so positively assert that this or that problem will never be solved by science.

Charles Darwin, *The Descent of Man*, 1870

Our genus *Homo* has only one living species, and one with remarkably little diversity at that. By contrast, our closest relatives, those in the genus *Pan*, boast two species – chimpanzees and bonobos – and four chimpanzee subspecies.

A hundred thousand years ago, the picture was very different. The genus *Homo* was extremely diverse, with species ranging from hulking *Homo erectus* to Hobbit-sized *Homo floresiensis*. How did we become so uniform?

That's the central enigma.

Waylaid by induction

Darwin gave us the answer to the origin of species. New species descend from their ancestors. Humans, no doubt, from apes. But which apes in particular? Darwin thought the great apes of Africa, chimps and gorillas. Alfred Wallace, the other architect of the theory of evolution, thought the great apes of Asia, the orangutans. He'd encountered them in the forests of Sarawak, and clearly they left an impression.[1]

For 100 years, it looked like Wallace was right. The missing links – Neanderthals, Java Man, Peking Man – were all found in Europe and Asia. It took till the 1970s to change the narrative. Omo 1, an ancient ancestor with far greater resemblance to us than any of the missing links, was unearthed near the Omo River in Ethiopia. In the 1980s, Kenya delivered Omo's possible ancestor in the form of 1.5-million-year-old Turkana Boy.[2] And the first forays into human DNA delivered evidence of an African Eve, mother of all modern humans. The very roots of our ancestral tree lay in Africa. Darwin was right after all.

But what about all those missing links in Europe and Asia?

Had we marched out of Africa – and terminated every other hominin in our way? Or had we interbred? Until 2010, we were held to be terminators. Then three Neanderthal ladies delivered up their DNA. And lo – they too were our ancestors. Since then, we've learned we carry snippets of DNA from other ancient ancestors: Denisovans and yet to be identified ghost species. (See box: "Ghost species".)

This chapter will make you giddy with its back and forth between dominant theories, a stark contrast with the century-strong steadiness of General Relativity. It provides a telling example of David Hume's 1739 warning about the perils of scientific induction: "There can be no demonstrative arguments to prove, that those instances, of which we have had no experience, resemble those, of which we have had experience." In other words, making predictions based on past observations and incomplete data is risky.[3]

Postcard from Bedrock

If you were to take a round-the-world trip 100,000 years ago, you'd meet some extremely colourful characters.

In Africa, you'd most likely bump into straitlaced *Homo sapiens*. But Europe and Asia would be a blast, dominated by the likes of hunky Neanderthals, hefty *Homo erectus* and the mysterious large-toothed Denisovans. Go south and the islands of Southeast Asia would delight you with diminutive *Homo floresiensis, Homo luzonensis* and gatecrashing Denisovans. The parties would be a riot, with revellers doing a lot more than just rubbing shoulders.

Take the same trip 30,000 years ago and the scene is quite dull. From Africa to Europe to Asia, it's all *Homo sapiens*.[4]

What happened? That's the central enigma of recent human evolution, and two major theories have vied to explain it.[5]

One held that all our colourful hominin relatives were dead ends on the family tree.[6] Our line traced back to a small army of modern humans who marched out of Africa and *replaced* all the others. In other words, we are a single-origin species. It fitted nicely with Darwin's evolutionary tree, in which new species arose from new branches, while extinct branches were last year's dead wood. This theory is referred to as "Out of Africa replacement".

The other theory held that we arose from interbreeding with these ancient relatives. In other words, we're a multiple origin species. It suggested our history was best represented by a trellis with intermingling tips. This theory is referred to as "Multiregional evolution".

The two theories have battled for five decades, first armed with fossils and stone tools, and in more recent times with DNA.

Since the beginning, anthropologists believed that Australia's Indigenous population would be central to solving the mystery of human origins.[7] Some of these anthropologists' ideas were fallacious, and the racist tone of their writing grates on modern ears. But they were right about divining the incredibly ancient history of Indigenous Australians. Nestled away for millennia in their Pacific bubble and unadulterated by the mass migrations that jumbled up the populations of the rest of the world, they represented "the very zero of civilization", according to one nineteenth-century writer.[8] "It is not too much to say that the untangling of the racial origins of the Australians is necessary for a successful reconstruction of the evolution of modern man," wrote Harvard anthropologist Joseph Birdsell in 1967.[9] The modern understanding of the origin of Australia's First Nations people has fulfilled Birdsell's prediction.[10] Genetic studies show that their roots go back to a group that made landfall on the southern continent at least 50,000 years ago. They represent the earliest known successful exodus of modern humans from Africa.[11]

Beginnings

To set the scene, we have to go back to 1859 and Charles Darwin's publication of *The Origin of Species*. The tome offered a theory to explain the "mystery of mysteries": where did new species come from? From their ancestors, claimed Darwin. All offspring are born with slightly different features. Natural selection favours the fittest. Over the eons, those selected differences add up to a new species.

Darwin was also the first to put forward the idea of a family tree for species, and he believed that a common ancestor lay at its base. [12]These days, we give that ancestor the name LUCA – the last universal common ancestor. Significantly, Darwin located all human beings on a single branch, putting paid to contemporary notions that races were different species.[13]

In painstaking detail (he'd slaved over it for twenty years), Darwin's *The Origin of Species* outlined the evidence for the gradual branching of species from their ancestral roots, using case studies on everything from barnacles to giant tortoises. But when it came to the origin of *Homo sapiens*, Darwin lost heart. As he admitted to Wallace, "I think I shall avoid the whole subject, as so surrounded with prejudices, though I freely admit that it is the highest and most interesting problem for the naturalist."[14] He decided to leave it for another day. "Much light will be thrown on the origin of man and his history," declared his closing paragraphs.[15]

It was "Darwin's bulldog", Thomas Henry Huxley, who famously took up the fight. Trained as a surgeon with a talent for anatomy – as a student he discovered "Huxley's layer" within the root of a human hair – he was also mesmerised by the mystery of mysteries. In 1849 he published his own studies on the changing anatomy of jellyfish species. "How extremely stupid not to have thought of that!" he blurted when Darwin finally articulated his theory. Then he became its most ferocious defender.[16]

Huxley was a formidable polemicist. In one famous debate, Samuel Wilberforce, Bishop of Oxford, baited Huxley by asking whether his descent from monkeys was on his mother's side or his father's. Huxley's riposte? He would rather have an ape for an ancestor than a bishop who distorted the truth.[17]

And then there was the "great hippocampus debate". Richard Owen, an eminent palaeontologist and Wilberforce ally, had shown that gorillas were missing a part of the human brain known as the hippocampus minor. Anatomy of course was just up Huxley's alley. He located the missing hippocampus minor in gorillas, dispatched Owen's claim and declared, "Life is too short to occupy oneself with the slaying of the slain more than once."[18]

Darwin took eleven years to enter the debate on human origins. It was in his 1871 publication *The Descent of Man* that he argued man's ancestors likely lay in Africa because of our resemblance to its upright walking apes – gorillas and chimpanzees.[19]

Wallace was more inclined to trace human origins to Asia. He had spent years collecting new species in the forests of the Malay Archipelago, particularly Sarawak on the northwest of the island of Borneo.[20] He saw kinship with its upright-walking gibbons and orangutans. Ernst Haeckel was even more confident about Asia as the home of the missing links – those protohumans who had branched away from the apes and evolved in a progressive sequence towards modern man.

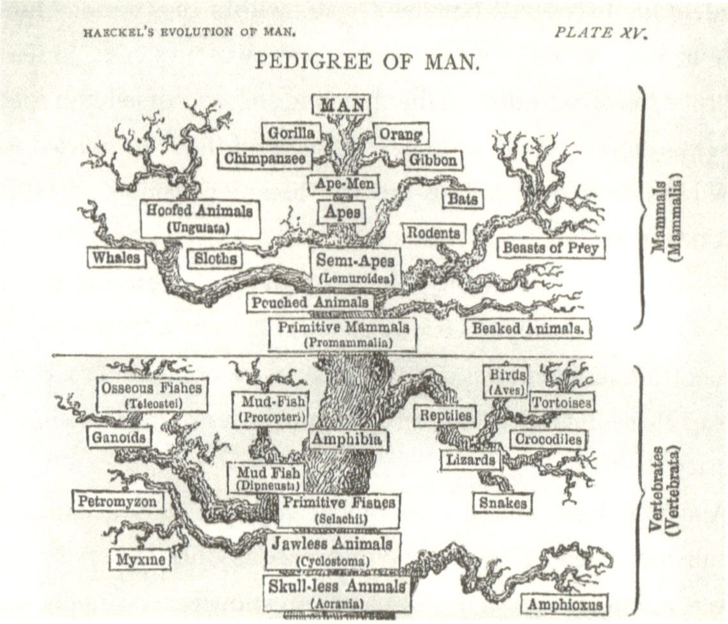

Figure 4. A detail from Ernst Haeckel's "Tree of Life", from *The Evolution of Man* (1879).

Haeckel, a medically trained zoologist, was another of Darwin's bulldogs. Unlike the pugnacious Huxley, he seduced the public with his breathtaking Art Nouveau–style paintings, collected in his 1899 book *Artforms in Nature*, and with his sketches of the embryonic stages of animals, from reptiles to humans. At the earliest stages, the embryos sported gill-like structures, supporting Darwin's view that all animals evolved from a common ancestor. As the embryos developed, they continued through forms that were related to their evolutionary ancestors: a fish stage, then a reptile stage, then a mammal stage, and so on – a sort of pageant of evolutionary history. As Haeckel famously put it, "ontogeny recapitulates phylogeny". Haeckel seems to have embellished the images to make his point. While human embryos do indeed have gill slits at an early stage, they do not recapitulate *every* stage of their evolutionary history.[21]

Until the 1970s, it looked like Haeckel and Wallace were right about Asia. African digs had only delivered a very distant relative: *Homo habilis*. Described by Louis Leakey in 1964, these metre-high hominins walked upright – a key criterion for inclusion in the human lineage.[22] They also made tools, busily banging rocks together and littering Tanzania's Olduvai Gorge with the flakes. But with their small brains, long arms and short legs, these two-million-year-old relatives lay on a very distant branch of the family tree. "In the 1970s, no anthropologist that I'm aware of thought of Africa as the source of modern human origins," recalls anthropologist Chris Stringer at the British Natural History Museum.[23]

So where were our direct ancestors – the missing links?

The idea of missing links referred to those protohumans who had branched away from the apes and evolved in a progressive sequence towards modern *Homo sapiens*. A famous 1965 graphic from *The March of Progress*, published by Time-Life Books in 1965, captured the idea in an enduring visual cliché captioned "The Road to Homo Sapiens".[24]

One of the first "links" was found in 1856 when quarry workers in Germany's Neander Valley ("Neandertal" in German) dug up a bow-legged skeleton with brow ridges.[25] At the time, he was mistaken for a rickety Cossack! In a seriously bad case of scientific induction, Germany's most famous pathologist, Rudolf Virchow, concluded that the short, bow-legged skeleton was that of a man afflicted by rickets. His colleague, palaeontologist August Mayer, concluded that the bony ridges above the man's eyes were caused by a habitually furrowed brow, the result of a painful life lived on horseback.[26]

By 1864, there was no more talk of rickety Cossacks. "Neanderthal Man" was proclaimed a missing link, *Homo neanderthalensis*.[27]

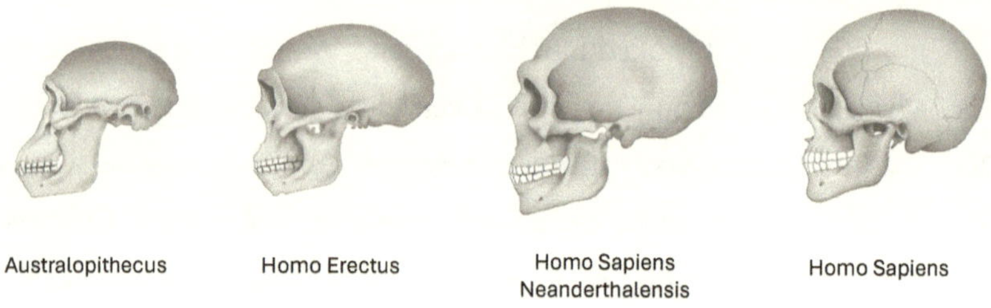

Figure 5. Skull features are crucial for identifying human ancestors. In archaic species, the brain case is long and flattened, brow ridges are prominent, the face protrudes, and the chin recedes. In *Homo sapiens* the brain case is globular, the brow ridges are gone, the face is flat, and there is a prominent chin.

To date, over 500 Neanderthal fossils have been discovered, with a range extending from the Middle East to Spain, France, Germany, Belgium, Croatia and even Siberia.[28] They roamed the continent from around 400,000 to 40,000 years ago. Generally stocky and short-statured with wide hips, it's a body built to survive the European ice ages. But it's their thick skull that provides the key to identification. It is elongated and flattened like a tortoise shell, with a forehead that slopes back. Above the eyes, like goggles, sit the bony brow ridges. The face and jaw protrude, the nose is large, and the chin recedes. And inside? Neanderthal brain capacity sometimes reached 1750 cubic centimetres, larger than the average 1350 cubic centimetres of modern humans. Overall anthropologists describe the Neanderthal skull as "robust".

By contrast, modern humans tend to be taller and more slender and have narrower hips. Our skull has a rounded, high dome like a US Second World War M1 helmet and is nearly as wide as it is long. Occasionally you might come across someone with brow ridges, but they're small by comparison with a Neanderthal's. Our faces are tucked under our skull, rather than protruding, and we have a pronounced chin. Overall, our skulls are thinner. In anthropologist-speak, if Neanderthal skulls are robust, ours are "gracile".

While quarry workers accidently stumbled on a missing link, Dutch doctor and anatomist Eugène Dubois deliberately set out to find one. Inspired by Haeckel's "sweeping, vivid and inspiring" prose and his belief that the missing links of human evolution would be found in Asia, Dubois joined the Dutch army and looked for opportunities to fossick in the Dutch East Indies (modern-day Indonesia).[29]

Sumatra disappointed, but in 1891 he hit the jackpot on the island of Java. He found a thick skullcap, wedged into the bank of the Solo River at a site called Trinil. Far more primitive than that of a Neanderthal, instead of brow ridges, it had a gorilla-like bar of bone (technically known as the supraorbital torus) protruding from above the eyes, a sloping forehead leading to a flattened tortoiseshell-shaped head, and a smallish, 900-cubic-centimetre brain capacity.

He named it *Pithecanthropus*, which means "ape man".[30] For Dubois, it was the missing link Haeckel had foretold. Alas, the scientific world failed to agree. Some considered it merely a giant ape. It left Dubois embittered, so much so that he hid the skull away in his home for twenty-three years. In time, more such skulls were discovered and Dubois' "Java Man" assumed pride of place as the defining example or "type specimen" of the newly discovered species *Homo erectus*.[31]

The banks of the Solo River turned out to be a treasure trove of *Homo erectus* skulls, perhaps because monsoon floods regularly swept away the unfortunate hominins. As their bodies decomposed, their dense limb bones sank, but their buoyant skulls snagged on the riverbanks. Sixty kilometres west of Trinil, at Sangiran, another riverbank delivered more than eighty *Homo erectus* fossils, including Java's oldest skull (an upper jawbone) dated to around 1.5 million years ago.[32] A third site, Ngandong, on a bend of the river some 70 kilometres to the northeast, offered up twelve of the youngest. They had brain capacities of 1100–1200 cubic centimetres, well within the range of modern humans. In 2020, geochronologist Kira Westaway at Sydney's Macquarie University, together with an international team, dated them to an astonishingly recent 108,000 years.[33]

That meant *Homo erectus* had roamed Java for at least 1.4 million years! And roam they did, over 5000 kilometres north, from Java to the outskirts of modern-day Beijing.

Peking Man

It was a "dragon bone" that gave him away. That's the term some Chinese use for the fossilised bones and teeth of extinct animals, because their size and shape is often monstrous. Ground into powders, they are considered to have remarkable medicinal value. So early twentieth-century European palaeontologists took to fossicking for fossils in Chinese apothecaries, sometimes with dramatic results. One dragon tooth testified to the existence of a new species of giant ape, *Gigantopithecus*. In 1921, Otto Zdansky, a newly minted Viennese palaeontologist, headed to the outskirts of Peking, Dragon Bone Hill or "Zhoukoudian", to dig up his own "dragon bones". He found a large molar with unusual cusps, and then a second. They did not belong to a giant ape.[34]

In 1926, Davidson Black, a Canadian physician and passionate anthropologist who headed the anatomy department of Peking Union Medical College, identified them as belonging to an ancient human. Then he did two audacious things. He submitted his findings on "two specimens of extraordinary interest" to the leading scientific journal *Nature*, where it was published a month later.[35] At the same time, Black participated in an event held for the visit of the king of Sweden – a patron of the Swedish geologist who led the excavations at Dragon Bone Hill. The event showcased the discovery of the teeth and announced that they belonged to a new missing link – "Peking Man".

It was all something of a publicity stunt. The Peking Union Medical College had been established by the US Rockefeller Foundation to train Chinese doctors in Western medicine, and the college boss initially took a dim view of Black's moonlighting with fossils. With the headline-grabbing news of "Peking Man", however, Black's work was suddenly a source of college prestige. He followed the headline by formally naming the species *Sinanthropus pekinensis* in the Chinese journal *Palaeontologia Sinica.*[36]

Proclaiming a new species, a missing link no less, based on two teeth was a risk! Critics immediately contested his analysis, suggesting the teeth instead belonged to a non-human carnivore. Fortunately for Black, his naming of a new human species was vindicated. Two years later, in 1929, Chinese anthropologist Pei Wenzhong discovered a complete skullcap that clearly belonged to an archaic human.

Peking Man, like Java Man, was deemed to be a *Homo erectus*. Overall, the site yielded six nearly complete skulls and the remains of forty individuals, some now known to date back to around 780,000 years ago.[37] Their cranial capacity averaged about 1000 cubic centimetres, with some individuals approaching 1300 cubic centimetres, well within the range of a modern human skull.

Anthropologists were riveted by Java Man, Peking Man and their kin. All the skulls sported the defining features of *Homo erectus* – they were generally "robust", with the bony bar above their eyes and a tortoiseshell-shaped skullcap. But there was also a great deal of variation. Some were less robust than others and showed features that seemed to be present in the modern populations of the region. How to explain this apparent march to modernity at Dragon Bone Hill? Franz Weidenreich – later described by a fellow scientist as "the flower of German civilisation and true culture" – came up with an answer.[38]

Physical anthropology, the study of human form, function and evolution, began in Germany in the mid-eighteenth century with studies of human skulls. Weidenreich was very much the "flower" of this tradition, with a particular interest in missing links. His mentor, Gustav Schwalbe, considered Neanderthals to be direct ancestors of human beings, and Weidenreich's university in Heidelberg was the home of the prized fossil *Homo heidelbergensis*, a European cousin of *Homo erectus*. With its larger brain and flatter face, it was considered by some to be the ancestor of Neanderthals.[39]

Weidenreich may have scaled the heights of German academia, but it was no bulwark against the Nazis. Weidenreich's science was weaponised to justify persecution of his own people. For him, it began in 1933, when, along with other Jewish German academics, Weidenreich was purged from the Heidelberg University. Unlike most German Jews, he went on to better things. When Davidson Black died suddenly from a heart attack, Weidenreich filled his spot at Peking Union Medical College and took over the study of the skulls of Dragon Bone Hill.

For Weidenreich, it was their variability that most intrigued him. Some were extremely archaic, with thick skulls, huge brow ridges and protruding faces. But others were finer, and their faces were flatter. Some of these also had intact incisors that revealed an interesting feature.

The rear upper surface was thickened, resembling an old-fashioned coal shovel.

Modern Asian people also had "shovel-shaped" incisors, which to Weidenreich suggested the trait had been passed down from *Homo erectus*.[40] (Geneticists now believe the trait arose from a more recent mutation.[41]) For Weidenreich, the modern-looking teeth and thinning of some of the skulls led him to assume he was seeing an evolutionary sequence. Peking man was evolving to resemble the modern people of the region. It was a sequence Weidenreich also discerned in the *Homo erectus* skulls of Java, and the Neanderthal skulls in Europe. Each ancient set of skulls, he came to believe, represented regional ancestors.[42]

Missing links down under

And improbable as it seemed, supposed missing links were also turning up at the bottom of the world, on a landmass that had always been separated from Eurasia by at least 100 kilometres of open sea.[43] A warning here: the following extracts from a leading scientific journal and a newspaper of record are grating to the modern ear, and, as we'll see, would be proven wrong.

In 1915, *Nature* described "the extraordinary features of Australia's earliest known human inhabitant, perhaps a member of the earliest family to ferry across Wallace's line and tramp dogs through Papua to Queensland".[44] The description was based on the skull of "Talgai Man", which was discovered on a cattle station some 125 kilometres southwest of Brisbane in 1886.[45] The *Nature* article continued:

> The canine tooth of the Talgai fossil was the largest human tooth so far discovered. The extremely primitive characteristics were such as to warrant one claiming for the skull considerable importance in anthropology, and worthy of being placed with such specimens as

> the prehistoric Heidelberg jaw and the Piltdown skull. There was … certain to be a great controversy …

The Wallace line referred to in the article lies between the islands of Bali and Lombok and marks the boundary between Asian fauna, such as tigers, and Australian fauna, such as tree kangaroos. A deep 35-kilometre strait churned by strong ocean currents, it remained a barrier even as the ice age lowered sea levels by 120 metres. The reference to the dog reflects the idea that Talgai Man might have been responsible for ferrying the Asiatic dingo to Australia. The Piltdown skull, meanwhile, was a forgery that misdirected anthropologists for decades. Found in a quarry in Piltdown, East Sussex in 1912, it appeared to be a prehistoric missing link, with a large human cranium and an ape-like jaw. In 1953, it was shown to have been made from a thoroughly modern human skull and an orangutan jaw. The Heidelberg jaw was a genuine fossil, an enormous ancient human jaw with teeth found in Germany in 1907 and now known as *Homo heidelbergensis*.

The New York Times upped the ante in 1926, when it reported the discovery of "Cohuna Man" … "the most archaic known".[46] He'd been unearthed in the town of Cohuna, some 270 kilometres north of Melbourne.[47] The *Times* continued, "all other human remains, including the Rhodesian, Piltdown and Talgai skulls, and the Neanderthal relics, are said to be 'comparatively modern' when put beside the Cohuna skull." Observing Cohuna Man's prominent brow ridge, receding forehead and elongated cranium, Weidenreich proposed a link between him and Java Man, both of whom he identified as ancestors of Australian Aboriginal peoples. For Weidenreich, the Australian finds completed a picture of "polycentric evolution" in which human evolution continued apace in four separate locations: western Asia, northeast Asia, Australasia and Africa. "We have no other choice but to abandon the generally accepted contention of there having been only one center of evolution from which, from time to time, new types would have sprung, and spread over the whole world, dislodging the older ones," he told a 1939 meeting of the American Anthropological Association in Chicago.[48]

It was a morally fraught moment for anthropologists. If there was no single centre of origin for humans, this could support older notions that racial differences ran deep. Indeed, in Darwin's day, different races were held to be different species.[49] In the nineteenth century, such racial theories were used to justify slavery. In the twentieth century, it was genocide. Weidenreich's family members were being shot, gassed or starved in concentration camps, including two of his daughters, who survived.[50]

Weidenreich was trying to make sense of what the Zhoukoudian skulls were telling him, but he was certainly alive to the racist dangers of his multiregional theory. That danger was made eminently clear by Carleton Coon, an anthropology professor at the University of Pennsylvania. Like Weidenreich, Coon believed that *Homo erectus* was the ancestor of modern humans. But their views parted ways when it came to race. For Weidenreich, racial differences were shallow, akin to those between dog breeds. For Coon, the differences went deep. He believed that races could be traced back to isolated populations of *Homo erectus* in different regions of the world that had evolved in parallel, like the arms of a candelabra. Later in his career, Coon proclaimed different human races to be "sub-species" and argued that mythical creatures such as the Sasquatch and Yeti were their ancestors – to which end he supported "Yeti-hunting" expeditions to Nepal and Tibet.[51]

Weidenreich, meanwhile, argued that *Homo erectus* populations were not isolated. Accomplished travellers, they had continued their peripatetic ways and occasionally interbred with distant relatives, just as modern humans do. The model Weidenreich proposed was not a candelabra but a "trellis", with multiple trunks and mingling branches. That trellis produced the mixed population that constituted the human race. As he wrote in 1939: "Crossing is not a late human acquisition which took place only when man had reached his modern phase, but must have been practiced ever since man began to evolve ... The races were all hybrids, and always had been."[52]

Weidenreich's theory was complicated. "He was trying to square the circle," says John Hawks, an anthropologist based at the University of Wisconsin–Madison. But that "squaring" was informed by the cutting-edge science of population genetics. Weidenreich was influenced by pioneering studies performed with fruit flies by Theodosius Dobzhansky, which showed how genes spread through fruit fly populations. The message Weidenreich took from Dobzhansky's fruit flies was to focus on populations, not on individual "types".[53]

As we'll see, the architecture of Weidenreich's trellis was wrong. Neanderthals are not the ancestors of Europeans, nor are Javan and Chinese *Homo erectus* the ancestors of populations in Indonesia, China or Australia. But his view that our species arose through the mixing of diverse populations was right.

As Hawks put it, "Weidenreich's trellis opened peoples' minds to an idea."

Race

Scientific theories of race have always been a lightning rod for moral and political debates. In Darwin's day, different races were widely assumed to be different species – a convenient excuse for slavery. Darwin's theory of evolution obliterated that excuse by arguing that all humans had descended from a common ancestor. Darwin himself was a fervent abolitionist, having witnessed the horrors of slavery during his voyage on the *Beagle*.[54]

But questions remained. How different were races? And how had these differences arisen? Race was a scientific knot that would take a century to untangle.

Weidenreich got it wrong in his view of regional ancestors. Neanderthals are not the ancestors of Europeans, nor are Javan and Chinese *Homo erectus* the ancestors of populations in Indonesia, China or Australia. But modern genomics has vindicated his idea of the mixing of populations and the shallowness of race. Take Europe, for instance. Its white population was once considered to be "Caucasian", one of the four supposed "major races". The other three were the "Mongoloids" of East Asia, the "Negroids" of sub-Saharan Africa and the "Australoids" of Australia and New Guinea. But Caucasians did not descend from a single white tribe. DNA blueprints from 10,000-year-old burial sites show that modern Europeans are the mixed descendants of four groups who occupied Europe back then: a western hunter-gatherer group, an eastern hunter-gatherer group, a farming group from the Near East, and another farming group from modern-day Iran. Each was genetically as different from each other as Europeans and Chinese are today. And they were hardly white. "Cheddar Man", a western hunter-gatherer whose remains were found in Cheddar in Somerset, had his DNA read in 2018. It revealed that he was dark-skinned, with dark hair and blue eyes. The average "white" British person inherits about 10 percent of their DNA from the likes of Cheddar Man.[55]

It is the same story for all supposed races: their apparent homogeneity – the similarities between Chinese people, or between Europeans – comes not from purity but from mixing. "There was never a single trunk population in the human past. It has been mixtures all the way down," noted David Reich, a Harvard geneticist and instigator of many of these studies.[56]

Humans also fail the technical definition of race, namely that to truly qualify as separate races, the difference between populations should be greater than the differences within them. Over 90 percent of the variations between humans occur within populations; less than 10 percent between. Overall, we are a species that shows little variety. One of our closest relatives, chimpanzees, show far more, and can be divided into at least three subspecies.[57]

The trellis down under

In 1965, Neil Macintosh, a charismatic University of Sydney anthropologist also known as "Black Mac", picked up where Weidenreich had left off. For him, the Australian ice age skulls bore "the mark of Java". His student Alan Thorne would take that epithet a whole lot further.

A one-time journalist at *The Sydney Morning Herald*, Thorne became hooked on the human origins story after meeting Macintosh. Trading his pen for a trowel, he began a PhD that involved analysing thirty skeletons from an ancient Aboriginal cemetery that had been accidentally disturbed by ditch diggers in 1962. The site, near the Murray River in northern Victoria and known as Kow Swamp, was in the same area where the sensational Cohuna skull had been unearthed four decades before.[58]

Thorne was convinced that these skeletons, which were dated at around 10,000 years old, did indeed bear the mark of Java. Like the handful of other ancient Australian skulls (which were about the same age), they were elongated, with prominent brow ridges, a protruding face and a receding forehead. In 1972, Thorne, together with geologist Phillip Macumber, published a paper on his findings in *Nature*. "The late

Pleistocene human remains from Kow Swamp," its opening summary declared, "display archaic cranial features which suggest the survival of *Homo erectus* in Australia until as recently as 10,000 years ago."[59]

Thorne was entertaining the idea that *Homo erectus* had either made the crossing to Australia themselves or interbred with a modern human population who did, perhaps in Java. But there was a major fact getting in the way of Thorne's story. If the features of *Homo erectus* were visible on 10,000-year-old skulls, they should be even more clearly etched on older skulls. However, when it came to Australia's most ancient skull, the "mark of Java" was nowhere to be seen.

Some 200 miles north of Kow Swamp lies Lake Mungo. You won't find a lake there. Instead, you'll be greeted by an eerie landscape where rampant erosion has chiselled the "Walls of China" from the red sandy clay, and naked white tree roots stand sentinel like ghosts. As geologist Jim Bowler discovered, these are not the only ghosts at Lake Mungo.

Previously a theologist, Bowler found his calling in reconstructing the history of this ancient landscape. As historian Billy Griffiths put it, "He was fascinated by this dry place that had been shaped by water, a landscape that could be read like scripture."[60]

Riding his motorbike across the eroding dune in July 1968, Bowler's attention was captured by some charred fragments of bone. His first thought was that they were the remains of a meal "burnt by early man". Many ancient campfires dotted the dunes. While no lake has graced Lake Mungo for 18,000 years, during the ice ages it was a bountiful oasis for Australia's first people, who roasted fish, shellfish and plentiful game.[61]

Bowler had to wait eight months for a team of archaeologists to inspect the find. In March 1969, Australian legends John Mulvaney (widely considered the father of Australian archaeology) and Rhys Jones arrived to dig out the burnt bones. When they found a human skull, the tenor of their conversation changed. This was not a meal burnt by early man; this *was* early man. The archaeologists packed the bundle of bones into Mulvaney's small brown vinyl suitcase, padded them with some of his clothes and sent them to Thorne's office at the Australian National University in Canberra.

The burnt bones were far more ancient than anyone suspected. Initially estimated at 32,000–25,000 years old, the final date came in at 40,000 years. Like a jigsaw puzzle, the skeleton had been shattered into hundreds of pieces, many of them coin-sized. It would take Thorne six months to put it back together again.[62]

When he was done, he encountered a paradox. Although it was by far the most ancient skeleton ever found in Australia, Thorne found himself gazing at the head of a thoroughly modern young woman. The gender and youth were suggested by her diminutive size, and modernity by the fact that she was not thick-boned but "gracile". Neither did she have any brow ridges to speak of; her forehead was high and her brain case was rounded.

Mungo Lady, as she was called, was not alone. Six years after she was unearthed, another late-afternoon bike ride through the dunes rewarded Bowler with the exposed skull of Mungo Man. Digging down revealed an intact male skeleton lying on his side, his hands carefully folded over his groin and daubed with ochre. Also estimated to be about 40,000 years old, Mungo Man's burial remains to this day one of the earliest ever known to involve the ritual use of ochre.[63]

Mungo Lady and Mungo Man were 30,000 years older than the Kow Swamp people and are the most ancient human remains ever to be discovered in Australia. If Australian Aboriginal peoples had interbred with Javan *Homo erectus,* as Thorne had hypothesised, then the "mark of Java" should be more evident here. It was not.

Thorne offered a resolution that fitted nicely with the ideas of Weidenreich. While the *Homo erectus* of Java were robust, China's Peking

Man was more delicately built, with a flatter face. In Thorne's view, if the Kow Swamp people bore the mark of Java, the Mungo people bore the mark of Peking. Thorne began articulating a theory of dual ancestry for Australia's first people.

In 1977, Thorne presented his ideas at a conference in Nairobi. Milford Wolpoff, from the University of Michigan, listened with interest. I spoke with Wolpoff, then eighty years old, by phone while he sheltered at home from the Covid-19 pandemic and a bitter Michigan winter. Of Thorne's presentation, he recalled: "I liked some of things he said. Not others. Everything was stimulating. It set me up for what happened next." It would be the beginning of a feisty but fruitful relationship.[64]

In 1979, Wolpoff got a chance to view the extraordinary variety of Asian *Homo erectus* for himself. US President Carter had formally recognised the People's Republic of China, and Wolpoff became one of the first American anthropologists to lay eyes on Peking Man. He also met and forged a friendship with anthropologist Wu Xinzhi at the Beijing Institute of Vertebrate Paleontology and Paleoanthropology, whom he described as "a gentle man, whose experience ranges from writing a book on gibbon anatomy to heading several expeditions to search for Yeti hairs ..."[65]

Next stop was Java and a visit to the laboratory of geologist Darwin Kadar. Kadar had possession of the most complete *Homo erectus* skull ever found, Sangiran 17, discovered in 1969 and dated to more than 750,000 years. While most of the skulls retrieved from the banks of the Solo River consisted only of the skullcap, the rock that preserved Sangiran 17 also preserved the bones underlying the face. When Wolpoff arrived, the facial bones were yet to be reattached to the skull. Kadar invited him to try.

With the pieces in a box cradled between his legs, Wolpoff reconstructed the skull from front to back using toothpicks and glue. When he turned the reconstructed skull side on and saw the protruding face, he was astonished to see something that looked quite different from Peking Man. "It didn't look like the Chinese fossils. Doggone it if it didn't look like an ancient Australian. How could Alan have been so right?"

Wolpoff's next stop was Canberra, where he wasted no time developing the film that carried the pictures of reconstructed Sangiran 17. He and Thorne began thrashing out their opposing views. "We thought so differently, but we got to the same place. It was a great collaboration," Wolpoff told me.[66]

In 1984, Wu, Wolpoff and Thorne minted their theory of multiregional evolution. Around the world, ancient human ancestors were evolving towards modernity. The *Homo erectus* of Java were giving rise to Southern Asians and to Australians; the *Homo erectus* of China were giving rise to East Asians; and Neanderthals were giving rise to Europeans.[67]

They credited Weidenreich's trellis theory. What they added was new data showing evidence for the mingled branches: fossil skulls with transitional features between archaic and modern from Australia, Java and China. They also folded in Thorne's "centre and edge" model: the global march towards modernity occurred as a result of mixing between populations, or "gene flow". But at the edges of the range, the mixing might be uneven, as the Aboriginal ice age skeletons showed. So while the march was universal, it did not occur in lockstep.[68]

They were also at pains to contrast their theory to Coon's "candelabra" model, which held that *Homo erectus* had splintered into separate lineages, ultimately giving rise to different races with deeply rooted differences.

Notwithstanding their efforts, many saw their multiregional evolution theory as supportive of racist theories. Wolpoff, like Weidenreich, is a Jew, and the misconstruing of his views was agonising for him and his wife, anthropologist Rachel Caspari. In 1997 they published a popular book to explain their stance, *Race and Human Evolution: A Fatal Attraction.*

A pivot to Africa, a drubbing for multiregionalism

Chris Stringer had other reasons to doubt multiregional evolution: the fossils were telling him a different story.

Today, Stringer is the highly revered Research Leader in Human Origins at the British Natural History Museum. In 1971, you might have mistaken him for a bearded, long-haired hippie, travelling around Europe in an old car. But rather than hanging out at rock festivals, Stringer was spending his time at museums. Kitted out with callipers, he was analysing the topography of Neanderthal skulls – measuring the angle at which the swept-back forehead met the top of the skull, recording their thickness and lengths, and noting curious embellishments such as a donut-sized bun on the rear or a yacht-like "keel" on the top. All in the service of his PhD thesis, which aimed to test the dominant paradigm: that Neanderthals were the direct ancestors of modern Europeans. His conclusion? They were not. Stringer had also run his callipers over a 195,000-year-old Ethiopian skull, Omo I.[69] This ancient African, with a robust yet rounded skull, was far more similar to modern humans than any Neanderthal he'd studied. It was also older than many of the European Neanderthals. For Stringer, Omo I was the more plausible ancestor to modern humans.[70]

Young Stringer bravely broke with the anthropological mainstream to argue that Africa was the birthplace of modern humans. This also made him one of the chief opponents of multiregional evolution. He had many comrades-in-arms. Stephen Jay Gould, the Harvard palaeontologist who entranced a generation with his luminous, literary tales of natural history, labelled multiregional evolution "a case of special pleading" for a "bizarre" theory:

> How could a new species evolve in lockstep parallelism from three ancestral populations spread over more than half the globe? Three groups, each moving in the same direction, and all still able to interbreed and constitute a single species after more than a million years of change?[71]

Among the most virulent adversaries of multiregional evolution was Australian palaeontologist Peter Brown. He was Alan Thorne's student. Like Thorne, he studied the anatomy of archaic Australian skeletons for his PhD. Unlike Thorne, he did *not* see a resemblance between the Kow Swamp skulls and Javan *Homo erectus*. "We didn't get along for decades," he told me.[72]

Instead, he saw evidence of a beautification practice: head binding. In one study, Brown compared three "robust" 12,000-year-old Australian skulls to Javanese *Homo erectus* skulls and skulls from the Arawe people of New Guinea, who practised head binding. His topographical analysis showed that angle for angle, the best match for the Australian skulls was to the Arawe head binders.[73]

Colin Groves did not see the mark of Java either. A fellow archaeology professor with Thorne at the Australian National University, he and his PhD student Michael Westaway took a computational approach to settle the argument over family ties. They selected twenty-seven features of the human skull that changed most during evolution and compared them between different archaic skulls. Rather than finding a family tie to *Homo erectus*, the computer linked the robust Australians of Lake Mungo to the hunter-gatherers of Africa and Israel. "The Mark of Ancient Java Is on none of Them", read the title of their 2009 paper.[74]

Professors Groves and Thorne disagreed, but they did so decorously. Indeed, they performed their debate on stages across Australia. Upon

Thorne's death in 2012, Groves penned a heartfelt obituary to a worthy adversary.[75]

Finally, there's Colin Pardoe, another of Thorne's former PhD students. Having interviewed Pardoe for many hours, I have decided he is "Talmudic". He can deal with conflicting things being true. A Canadian, he was drawn to Australia in 1980 by three things: "incredible beaches, incredible bikinis and Cohuna Man". His PhD thesis aimed to explain "Prehistoric human morphological variation in Australia". Why was it that people who lived along the Murray River were robust, while those who lived on the shores of Lake Mungo were gracile? The answer he found was not due to the "mark of Java" or the "mark of Peking" but to well-understood evolutionary forces at work across the continent. Natural selection, for instance. All over the world, it shapes people to fit their environment.[76] Like the lanky Masai, desert people everywhere evolve to be tall and thin. According to Pardoe, the people who lived on the shores of Lake Mungo were desert people taking advantage of the occasional filling of the lake. The Murray people, by contrast were living on a reliably resource-rich river floodplain that even today serves as "Australia's breadbasket". No surprise, then, that they were generally more robust. It still holds true, says Pardoe. He points to the fact that hunky rugby players tend to be "Murray people" while lean, tall Aussie Rules players are typically "desert people".

It's not just natural selection that explains diversity in populations. Natural selection's less sexy cousin – genetic drift – is also a major force. Genetic drift refers to the phenomenon that in small populations, random mutations, whether useful or not, are more likely to gain a foothold. Along the Murray, marital law meant that partners were sourced from the next tribe along the river. Pardoe documented how this traditional practice concentrated quirky traits among groups along the Murray. Just as a funny earlobe might be shared by an uncle and his great nephew, odd bumps on the skull tend to run in these groups. Pardoe has used those family bumps to repatriate skeletons from museum collections to their ancestral lands. He also found that an elongated head shape clusters around the Murray. As distance from the Murray increased, heads grew rounder, suggesting that this feature, like the quirky bumps, concentrated in the region through genetic drift.

Yet Pardoe didn't *entirely* disagree with Thorne. He admits that when he looks at the skulls that represent Java Man, he does see an "Australian look" – much the same look that Weidenreich and Wolpoff saw. As Pardoe describes it for the Aboriginal communities he works with, skulls with an Australian look have a:

> long and narrow head – hence the difference in blocks for the squatter's hat and the stockman's hat. The squatter had an Irish head shape. The stockman had a Pilbara Aboriginal head shape.[77]

Throughout the 1970s and 1980s, all these theories about Australian origins competed in the brutal arena of scientific debate: were Mungo Lady and Mungo Man best understood as the result of *Homo erectus* ancestry, African hunter-gatherer ancestry, head binding or local evolutionary forces? But for all the brawling, no one could score a knock-out punch. Then, along came DNA.

All about Eve

Like a volcanic upthrust, a single paper published in *Nature* in 1987 forever reshaped the landscape of the debate.[78]

Allan Wilson, holder of a MacArthur Fellowship or "genius grant", with his colleagues Rebecca Cann and Mark Stoneking at the University of California, Berkeley, analysed the partial DNA code of representatives of five populations – from Africa, Europe, Asia, New Guinea and

Australia – and discovered a striking thing. There was relatively little genetic difference between these populations. The greatest genetic diversity existed between populations inside Africa.

What does diversity mean when it comes to DNA? Along the human DNA code, at any given position, you can find variations. We can think of them as letter substitutions, analogous to the way that some English words have alternative spellings, such as grey/gray, colour/color, aluminium/aluminum. DNA diversity refers to how many alternative spellings are identified. Across African populations, Wilson's team found many of these variations. Outside Africa, there were relatively few.

It was a striking parallel to what is observed in human languages. Notwithstanding the difference between Italian, Hindi and Russian languages, they all derive from a common Indo-European language family. Overall, European languages trace to just two language families: Indo-European and Uralic. Africa boasts at least six language families: Afro-Asiatic, Niger-Congo, Nilo-Saharan, Austronesian, Khoe-Kwadi and Khoisan.[79] The simplest inference of Wilson's finding was that the modern human family originated in Africa and had been there for a long time – long enough to diversify into different tribes with diverse DNA and diverse languages. Eventually, one tribe left to colonise the rest of the world. In other words, our family tree was rooted in Africa, where multiple branches of the family had spread out. By contrast, the rest of the world had been populated by a single offshoot.

Wilson's lab did not read the entire three-billion-letter DNA blueprint that definitively identifies every human being. That wasn't possible in the 1980s. Instead, they analysed the variations found in a shorter, 16,000-letter code. While the complete DNA blueprint is vaulted away in the nucleus of each human cell, this shorter one is carried by mitochondria, the power stations of the cell that generate energy from sugars. Tiny and torpedo-shaped, about a thousand mitochondria exist in each cell. They resemble bacteria, and once upon a time they probably were, given they carry their own DNA. More than a billion years ago, they began cohabitating with a much larger but energetically challenged type of cell.

Mitochondrial DNA, like all DNA, can also be used as a clock. This is because variations in the DNA (mutations) accrue steadily over time. So by comparing the average number of mutations in the DNA of two populations, it is possible to estimate how long they have been separated. Mitochondrial DNA also exclusively traces the maternal line. That's because sperm leave their own mitochondria behind when they burrow into an egg.

Mitochondrial DNA and African Eve

A human egg is supplied with a payload of mitochondria from the mother. When a sperm fertilises that egg, it leaves its own mitochondria behind. So mitochondrial DNA only comes from mothers; there is no paternal contribution. However, not all females go on to have fertile female offspring. For a woman to keep passing on her mitochondrial DNA, her daughters must have daughters and so on. So the likelihood is that in any population, the mitochondrial DNA of most women eventually reaches a dead end. For the rare woman whose daughters keep having daughters, her mitochondrial DNA will come to dominate the population. This explains why Allan Wilson found that all living women can trace their mitochondrial DNA to an Eve who lived in Africa around 200,000 years ago. After 10,000 generations a single female appears to have won the lottery.

A similar phenomenon explains how a village may end up with a predominance of a single surname. Because the surname, like the

Y-chromosome, travels with the male line, the surnames of those rare males whose sons have continued to father sons will dominate the population.

Using mitochondrial DNA, Wilson placed African populations at the root of a tree that dated to about 200,000 years ago. The rest of the world had been colonised by a single twig that sprouted around 60,000 years ago. Wilson's analysis even identified what appeared to be the African Eve, the maternal ancestor of the colonists who marched out of Africa, replacing every other *Homo* species in their way. (See box: "Mitochondrial DNA and African Eve".)

If Wilson's clock was correct, it sounded the death knell for the trellis theory of human evolution. That trellis had roots in two-million-year-old *Homo erectus*, and its arms were entwined with 400,000-year-old Neanderthals. Wilson's tree summarily lopped off both of them.

And then came what seemed like the knock-out punch. A year after Wilson's mitochondrial DNA study placed the root of the human tree in Africa, Chris Stringer and his colleague Peter Andrews at the British Natural History Museum backed the DNA evidence with a mass of fossil evidence to argue for a singular recent African origin, rather than a slow and gradual evolution across the planet from archaic ancestors.

Stringer had already argued that Ethiopia's Omo I had better credentials to be the ancestor of Europeans than Neanderthals. The 1988 paper in *Science* by Stringer and Andrews rooted the tree of man even further back in Africa, among Morocco's Jebel Irhoud skulls (now known to be about 300,000 years old), which showed a mix of primitive and modern features.[80] They also argued for the implausibility of multiregional evolution and "gene flow" to keep archaic human populations blended: "it seems improbable that it could prevent increasing differentiation in a very widely distributed species over long periods of time." The paper ended with a parting shot: "paleoanthropologists who ignore the increasing wealth of genetic data on human population relationships will do so at their peril."

All in all, it was a declaration of victory for the Out of Africa theory. Yet it was not a knock-out punch.

The multiregionalists fight back

"The 1990s were an era of intense debate", recalls John Hawks.

Wolpoff and Thorne did not accept that mitochondrial DNA had disproven their theory. For one thing, they claimed, mitochondrial DNA could not tell the whole story, since it is only carried by the female line. The male line, traced by the Y chromosome, might yield quite another tale.[81]

Others queried the reliability of the mitochondrial DNA clock, which had rooted our origins to 200,000 years ago and lopped Neanderthals and *Homo erectus* off our family tree. That clock assumed the same tick-rate for all DNA. But researchers who focused on nuclear DNA measured different timing. For instance, one group reported that particular genes carried by nuclear DNA traced back to an ancestor who lived not 200,000 years ago but a million years ago! One such gene, microcephalin, had been linked to the ballooning of the human brain.[82]

Computer scientists also challenged the way Wilson had drawn the family tree. Using Wilson's own data, they showed that you could potentially draw thousands of different trees. And the trees were just as likely to be rooted *outside* of Africa as inside. "We're not saying ... that [the origin] is definitely non-African, but rather that you can't tell," said Harvard anthropologist Maryellen Ruvolo.[83] Other argued that the data did not support the drawing of a tree at all. Alan Templeton, a geneticist and ecologist from Washington University, used a statistical approach to test whether the DNA of different species fit a family tree structure. When he applied the test to African elephants or chimpanzees, he could draw trees. On the other hand, the DNA of moray eels, which slither halfway across the planet in the Indo-Pacific Ocean, showed no tree structure. Humans are more like eels, says Templeton. "One of the unique things about humans is their amazing ability to disperse much more than other land mammals," he noted. Templeton's analysis found that human genomes going back 650,000 years sat better on a trellis than a tree.[84]

And so the boxing match between multiregional origins and Out of Africa continued into the 1990s. But the fast-developing technology of DNA sequencing promised to finally settle things. In 2001, reading an entire nuclear DNA blueprint (a genome) cost $3 billion and took over a decade. But in the early years of the twenty-first century, the speed and cost plummeted. By 2007, Wilson's successors had access to a thousand human genomes. They could ask afresh: was the human population best represented by a trellis or a tree?

Most of them drew trees, and most were rooted in Africa 50–70,000 years ago.[85] The big data also laid bare the remarkable genetic similarity between all humans. Whether Inuit or Aboriginal Australian, we are 99.9 percent the same. With that degree of sameness, the species had to have a single origin. Surely the theory of multiregional evolution was now well and truly debunked?

Sorry – but no. We have another big round to go in this boxing match.

Neanderthals in the family tree

In 2010, Svante Pääbo turned the theory of human origins upside down. Again.

A twenty-five-year quest to reveal the secrets of ancient people by decoding their DNA succeeded beyond his wildest dreams when his team at the Max Planck Institute read the first Neanderthal genome. It was the stuff of science fiction. But the greatest revelation was not about Neanderthals. It was about us. We were not a single-origin species; we were part Neanderthal.

Blame it on the Swedish-born scientist's obsession with antiquity.[86] As a thirteen-year-old, he participated in digs in Egypt with his mother. He tired of sorting pottery, but he had caught the antiquarian bug. He completed a medical degree at the Uppsala University and began a PhD to study the DNA of an adenovirus. All the while, he attended lectures at the Institute of Egyptology and took classes in Coptic, the language of Pharaonic Egypt spoken during the Christian era. The combination of DNA and Egyptology fired up his mind. Might it be possible to extract DNA from a mummy? Forget pottery or Coptic texts; this ancient DNA text could crack the mysteries of the Egyptian dynasties. Pääbo secretly began beseeching museum curators for mummy flesh. Receiving a response from the Bode Museum in communist East Berlin, he took a ferry across the Baltic Sea to visit the grim building whose roof still bore huge holes from bombs and artillery shells. He took biopsies from over thirty mummies and, back in his Swedish lab, attempted to extract DNA. A 2400-year-old child delivered. Pääbo's groundbreaking paper was published in *Nature* in 1985 with a cartoon drawing of a mummy on the cover. Its unravelling shroud consisted of threads of DNA.[87]

The fame of that paper catapulted Pääbo to the lab of another brilliant maverick: Allan Wilson in Berkeley. Here Paabo carried out further fantastic feats, extracting the mitochondrial DNA of the extinct Tasmanian thylacine, commonly known as the Tasmanian tiger because of its stripes. Taxonomists had suggested it was most closely related to South American marsupial wolves, but Pääbo's DNA testing showed it was more closely related to Australian marsupials like the Tasmanian devil.[88]

Extinct animals, however, were a diversion; Pääbo was itching to get back to human history, and to address a growing concern. He'd begun to suspect that the DNA he'd retrieved from the Bode Museum may not have been that of a mummy, but of a curator. Most studies were showing that ancient DNA was usually shredded into pieces about 150 letters long, but the mummy DNA he'd retrieved was thousands of letters long.[89]

As the glow of his mummy triumph began to fade, the indomitable Pääbo channelled his energy into developing a protocol that would eradicate so much as a whiff of contaminant DNA. Masked, gowned and gloved technicians would work in clean rooms sterilised by UV light and aerated with filtered air. Only here, equipped with sterile dental drills, would they bore into ancient bones to remove the fine powder from which they would extract DNA.

The painstaking techniques paid off. In 1994, Pääbo's lab, now at the University of Munich, was entrusted to analyse the DNA of Ötzi, the 5000-year-old ice man who'd thawed out of a glacier in the Tyrolean Mountains three years before. An analysis of Ötzi's mitochondrial DNA showed that he was closely related to current-day central and northern European populations.[90]

It was a great success, but Pääbo's sights were already set on a human far, far more ancient than Ötzi.

In 1997, Pääbo's lab gained access to a bone fragment from the world's most prized Neanderthal, the "type specimen" unearthed in a cave in the Neader Valley in 1856. They successfully isolated mitochondrial DNA. It was utterly unlike that of modern humans. It was also evidence that Pääbo's mentor Allan Wilson was right. Modern humans had a single African origin. They did not mix with Neanderthals.[91]

It would take thirteen more years to read the nuclear genome of a Neanderthal. If reading the first modern human genome had been a Herculean feat, a Neanderthal genome was a whole other level of pain.

Those Neanderthal bones that occasionally yielded mitochondrial DNA never yielded nuclear DNA. There was a simple reason: each cell carries around a thousand copies of its short mitochondrial DNA code but only two copies of its unfathomably long string of nuclear code. Extracting nuclear DNA was like mining some vanishingly rare ore. Pääbo and his team relentlessly refined their techniques: from finding ways to rid their samples of contaminant bacterial DNA to trialling methods that required just one tenth the amount of DNA to yield meaningful results. Finally, Pääbo's team were able to piece together a full nuclear genome by combining DNA from bone fragments of three 38,000-year-old female Neanderthals from Vindija cave in Croatia. Harvard population geneticist David Reich compared that genome to that of a modern human. His analysis stunned the world. Modern humans carried Neanderthal DNA.[92]

Pääbo was more stunned than anyone: "It contradicted the strict Out of Africa hypothesis, of which my mentor Allan Wilson had been one of the main architects. It contradicted what I myself had believed to be true. Neanderthals weren't totally extinct. Their DNA lived on in people today."[93]

As well as redrawing the modern human family tree, the Neanderthal DNA delivered a chronicle of their life and times. Neanderthals branched away from modern humans about 500,000 years ago. They were highly inbred, suggesting they lived in small, isolated groups, a factor that likely contributed to their demise. Their DNA also gave us the first clues as to what Neanderthals were like in the flesh. Some carried genes for fair hair and a variety of keratin genes that might have toughened their skin against the cold. And we got a partial answer to a long-standing question: did Neanderthals have the capacity for language? In modern humans, the gene crucial for language is known as FOXP2. People who carry mutations in this gene are intellectually normal but have difficulty speaking. Chimpanzees and gorillas carry a different version of the FOXP2 gene to humans. And Neanderthals? They carried largely the same version of the gene as us, albeit with some changes to the way it performs.[94]

Overall, Neanderthal DNA was 99.7 percent similar to our own, but all along the code there were recognisable differences. Imagine human DNA is like modern English, while Neanderthal DNA is like the Shakespearean version. Neanderthal DNA might read "But soft! what light through yonder window breaks?", while the modern text reads "Hey, what's that light shining through the distant window?"

These quaint alterations of letters and words, technically known as single nucleotide polymorphisms or SNPs, allowed geneticists to search modern genomes for Neanderthal idioms. They found them. Outside of Africa, the DNA text of every person on the planet is peppered with about 2 percent Neanderthal. Different chunks have been inherited by different people, so that overall, around 30 percent of the Neanderthal genome is represented in a modern human genome.[95] Africans were long thought to have no Neanderthal DNA, but an analysis in January 2020 showed that they carry around 0.6% Neanderthal DNA, probably a result of back migration of populations that had previously ventured into Eurasia. What this suggests is that as modern humans left Africa, they encountered and bred with Neanderthals. Their trysts have been dated to around 50,000 years ago.[96]

The fact that we kept 30 percent of their DNA (in total across our species) and rejected the rest suggests that some Neanderthal DNA was detrimental to our fitness. Neanderthal DNA is conspicuously absent from genes related to male fertility and genes on the X chromosome. That's a pattern seen in species that are right at the limit of biological compatibility.[97]

Some of the Neanderthal genes we retained have been beneficial.

Many have to do with skin and hair; these adaptations may have been useful for dark-skinned people coming out of Africa. Others beef up the immune system. Neanderthals may have evolved these genes to protect themselves from infectious diseases as they huddled up in their freezing caves. When the Covid-19 pandemic hit, some people developed mild infections; others were severely affected. It turns out that genes inherited from Neanderthals significantly determined a person's risk.[98]

Denisovan DNA

2010 turned out to be a year of wonders.

Six thousand kilometres to the east of Croatia's Vindija Cave, at the foot of Siberia's Altai Mountains, lies some prime real estate. Denisova Cave is large and well lit, with a skylight that doubles as a chimney for cold winter nights. At the crossroads of Siberia, China, Mongolia and Kazakhstan, it offers a commanding view of the mountains and verdant river valley below.

Digs in this cave have delivered stone tools and ornaments crafted from animal bones, teeth, mammoth ivory, ostrich eggshell and a beautifully carved stone bracelet. But who were the artisans? Surely the occupants of this stone-age penthouse left behind a skull or two as deeds of ownership.

They did not.

Hyenas had done their worst; the only thing captured in the archaeologists' sieves were tiny fragments of bone, too tiny to determine what sort of creature they came from. All the archaeologists could do was speculate. Given the sophistication of the handwork, they assumed the occupants were of our kind. And so things would have stood, if not for Svante Pääbo.

Amid the rubble of tiny bone fragments, one raisin-sized piece stood out. Expert anthropologists identified it as the tip of the fifth finger of a right hand, from a hand that belonged to our family group *Homo*. But which *Homo*? The pinky fingertip was tiny, but it was rich in DNA. Using their meticulous methods, the Paabo team delivered a coplete genome. Once again, David Reich's analysis stunned the world.[99]

The pinky bone belonged to a young woman who lived 50–30,000 years ago. She was neither a *Homo sapiens* nor a Neanderthal. She was the first known member of a new group. But unlike every other *Homo* in the fossil record, her kind had not been discovered through a fossil skull, but through fossil DNA. They were named Denisovans. The jury is still out as to whether they should be considered a different species, a subspecies or one of our own. (See box: "Species or subspecies?")

The mysterious young woman wasn't alone. A hefty tooth also yielded Denisovan DNA. It was so large and weirdly shaped that archaeologist Bence Viola, who found the tooth, initially attributed it to a cave bear.[100]

Neanderthals were at Denisova Cave too. A toe-bone's DNA yielded the ID. It belonged to a female whose parents had likely been half brother and sister. It was a glimpse into their unforgiving world. Hunkered down in the cave to survive the ferocious winters, their numbers were dwindling. Another 90,000-year-old female whose ID came from a bone fragment turned out to be a hybrid. Named Denny, her mother was a Neanderthal (evidenced by her Neanderthal mitochondrial DNA), her father a Denisovan.[101]

The teeth and bone fragments of Denisova Cave had delivered a DNA bonanza. But in this cave, even the dirt delivered DNA. (See box: "Dirty DNA".) And that DNA revealed evidence of our kind. All up, the soil layers showed the cave was a drawcard for three kinds of humans.[102]

For Russian and Chinese anthropologists, several of whom champion models of multiregional evolution – which they refer to as a "braided stream" rather than a "trellis" – the cave was just the place where this kind of evolution through interbreeding took place.[103] The era of DNA archaeology had begun, and things were only about to get even more mysterious.

Dirty DNA

Since 2003, scientists have been able to directly retrieve DNA from dirt – identifying the signatures of everything from ancient mosses to woolly mammoths, with some dating back 400,000 years. DNA ends up in the soil in a number of ways. It might be from a disintegrating bone, from the cellular soup oozing out of a decomposing animal, or from a live creature's urine and faeces. Clays are very good at binding up DNA. While the vast majority of soil DNA is microbial, researchers can fish out the DNA of the species they want using lures of matching DNA. They also have tests to ensure that the source of the DNA is ancient rather than the modern DNA of the scientists handling the samples. In ancient DNA, the letter cytosine is often chemically altered to the letter thymine.

Once researchers had a high quality Denisovan genome, they could attempt the same sort of "search" they had performed for Neanderthals. Were there any matching Denisovan words or phrases in modern human DNA?

There were. In a population as far from Siberia as one could imagine: Australian Aboriginal peoples, and Papua New Guineans. As early anthropologists realised, these populations had an ancient history. Going back 50,000 years, their ancestors lived on the shared ice age continent known as Sahul.[104] The DNA of their descendants was up to 4 percent Denisovan. The populations of Europe and Asia carried less than 0.1 percent. So while Neanderthals left their mark evenly on the emigrants from Africa, the Denisovan mark was skewed to the descendants of the settlers of Sahul.[105]

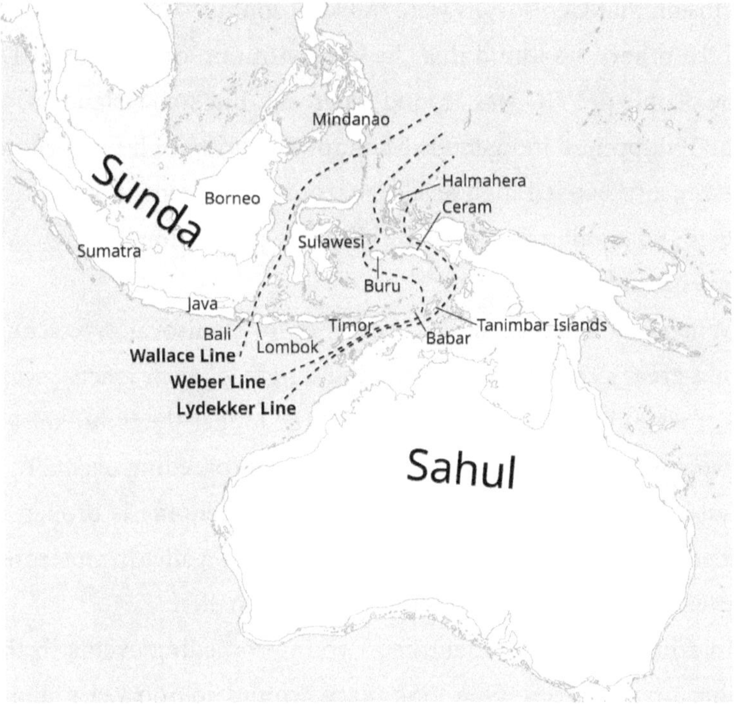

Figure 6. During the ice ages, low sea levels created the landmasses known as Sunda and Sahul. The islands in between, surrounded by deep seas, are known as Wallacea, after Alfred Wallace. Getting from Sunda to Sahul has always required crossing about 100 kilometres of open sea. The Wallace, Weber and Lydekker lines demarcate dramatic differences in the make-up of plant and animal species.

How to explain it? One possibility is that Denisovans trekked southward to Sahul, where they later interbred with modern humans!

That's the interpretation favoured by Murray Cox, a computational biologist at Massey University in Auckland, New Zealand. If archaeologists add chapters to prehistory by digging through cave dirt, computational biologists do it by digging through DNA. After sifting through the DNA of people who live on the islands that lie like stepping stones between the Asian mainland and Sahul, Cox's group reached the conclusion that Denisovans were trekking south.[106]

Like others, he found that the concentration of Denisovan DNA in the people of PNG was around 4 percent. But some islanders, such as the Philippines' indigenous Mamanwa, who have hunter-gatherer ancestry, also carried high levels, at around 2 percent. A more recent study found another indigenous Philippine population, the Ayta Magbukon, has close to 5 percent.[107]

When Cox searches modern genomes for Denisovan tracts, he can learn a great deal. For one thing, the length of such tracts gives an idea of when interbreeding happened. The longer the unbroken tract of Denisovan DNA, the more recent the interbreeding event. This is because with each generation, each parent's genome is broken and shuffled like a pack of cards. With each generation after the interbreeding event, the length of this chunk becomes smaller.

In 2019, Cox's team identified two interbreeding events in these southern populations. One took place around 50,000 years ago; the other a mere 30,000 years ago.[108] Given that the Denisovans' partners were already in Sahul 50,000 years ago, Cox suggested Denisovans might also have made it to Sahul. Regardless of where exactly the trysts took place, the surprise finding suggests that Denisovans survived in this part of the world until 30,000 years ago, longer than Neanderthals or any other archaic population. Surely the Denisovans must have left some remains.

Species or subspecies?

Researchers can't agree whether Neanderthals and Denisovans are the same species as us, or a subspecies. The idea of a species seems not to be as simple as whether or not they can interbreed – the biological definition employed by influential evolutionary biologist Ernst Mayr. This debate gets very messy.

Zoologist and director of the Australian Museum Research Institute Kris Helgen points out that Mayr's definition fails all too easily. Polar bears and grizzlies, for instance are different species and are usually kept apart, because polar bears live on sea ice while grizzlies live on land. But they can interbreed, and more and more hybrids are turning up in the wilds of Canada and Alaska as the sea ice retreats and polar bears stray onto the tundra. Even lions and tigers are not clear-cut. Now that we can read their genomes, we can see there was a period in which there was interbreeding after their species split.

It seems we've all been too precious with our idea of species. Darwin was more pragmatic. His view was, "a species is what a competent taxonomist says it is".[109]

Show us the body (of a Denisovan)

Like a detective's pinboard with identikit sketches, fingerprints, newspaper clippings and maps, all festooned with strings making tentative connections between them, the search for the missing Denisovans has for fifteen years remained a cold case. In that time, tantalising new clues have been added to the board. In 2016, Denisova Cave yielded another piece of a Denisovan, identified by its DNA. It was the top of an extremely thick skull; some archaeologists thought it resembled that of a *Homo erectus*, but the fragment was too small to be sure.[110]

In 2019, 2800 kilometres south of Denisova Cave, in the high Tibetan Plateau, a much bigger piece surfaced. A Tibetan monk first spied it thirty-nine years ago. Partial to meditating in Baishiya Karst Cave, another prime piece of Palaeolithic real estate 3200 metres above sea level, his tranquillity was disturbed by a big jawbone with two giant molars, lying in a corner of the cave. The monk gave it to his teacher, the Sixth Gung-Thang Living Buddha, who donated it to Lanzhou University. The university archaeologists had no idea how to classify the jaw. It sat on a shelf collecting dust until, four decades later, the big jaw with oversized teeth piqued the interest of archaeologist Dongju Zhang.

Perhaps she'd seen an article in *Nature* in 2014, showing that Tibetans adapted to high altitudes thanks to a gene inherited from Denisovans.[111] She sent an email to Jean-Jacques Hublin, director of the Department of Human Evolution at the Max Planck Institute for Evolutionary Anthropology in Leipzig. "At the sight of the pictures on my screen, my heart jumped," recalled Hublin.[112] Hublin and Zhang decided to collaborate to try to identify just what type of jaw it was. While the molars no longer preserved DNA, their dentine preserved fragments of collagen, the hardy protein that comprises the framework of teeth. It was enough for an ID: the 160,000-year-old jaw was indeed Denisovan.[113] A year later, the very sediments of the Tibetan cave yielded Denisovan DNA, confirming it had been a Denisovan stronghold from 100,000–45,000 years ago.[114]

Sleuthing Southeast Asia

The mystery of the Denisovans is mesmerising the archaeological world. New papers seem to appear weekly. Archaeologists, anthropologists and geneticists are pinning more clues to the detective board.

And the clues are leading them south. Recall Murray Cox's analysis of the DNA of isolated populations who live on the islands between Asia and Papua New Guinea. It suggested that Denisovans had been in this part of the world as recently as 30,000 years ago. So which archaic fossils from Southeast Asia from around this period might be the missing Denisovan? Allowing for the rubberiness of the time estimate, we have three suspects in the region. (See box: "Island hobbits".)

Two of them are just too weird to consider, in my humble opinion. These are the dwarf hominins found recently on the islands of Flores and Luzon. They're around a metre tall with brains about a third the size of ours, and a suite of extremely primitive features. They resemble the east African archaic hominins like the Australopithecine Lucy and *Homo habilis*. Just how they got to these Southeast Asian islands is an enigma. But these diminutive hominins are hardly likely to be Denisovans. For one thing, the fossil evidence says Denisovans were hefty with bear-sized teeth.

And that leaves *Homo erectus*. In many ways, he fits the picture. His range extends across Europe and Asia, and he's most populous on Java – the kick-off point to the islands of Southeast Asia and Sahul. Where he falls down is that many archaeologists view him as too archaic to be a candidate for a Denisovan. He sits on an evolutionary branch that split from our own over 2.5 million years ago. By contrast, Denisovans and Neanderthals split from our branch a mere 500–700,000 years ago. Researchers such as Joao Teixeira at the University of Adelaide have searched for traces of two-million-year-old DNA in people from Southeast Asia, and were not convinced.[115]

Island hobbits

Homo floresiensis lived on the island of Flores in Indonesia as recently as 60,000 years ago. The oldest fossils date back to more than 100,000 years, but based on the dates of stone tools found elsewhere on the island, they may have been present for 840,000 years. The fossils, which were found at Flores' Liang Bua Cave, include a largely complete skeleton and skull, as well as parts of eleven other individuals, and show they were about a metre tall, walked upright and had a brain capacity of some 425 cubic centimetres, about one third that of a modern human.

Homo luzonensis lived on the island of Luzon in the Philippines. Fossils found here at Callao Cave include a few curved toe bones, seven teeth, two finger bones and part of a femur. They tell us that this hominid existed over a similar time span to *Homo floresiensis.*

Both species are extremely challenging to place in the human family tree. *Homo floresiensis* shows some traits last seen in African ancestors such as *Homo habilis*, while *Homo luzonensis* recalls the even more primitive Australopithecines such as "Lucy". It's hard to imagine Lucy or *Homo habilis* trekking and sailing to Southeast Asia from east Africa over two million years ago. A more tenable theory is that some of Java's *Homo erectus* managed to sail or raft across – and then shrank. We know this happened on Flores with stegodons, archaic members of the elephant family. It's a well-known phenomenon called island dwarfism. Cases also include dwarf hippos that shrank to a tenth the mass of their mainland ancestors. Island gigantism occurs too, as evidenced by 'rodents of unusual size' that increased their mass 100-fold.[116] "Evolution goes crazy on islands," notes Kris Helgen, director of the Australian Museum Research Institute.[117]

It can happen fast. Small founding populations and extreme conditions fire up the evolutionary engine, so that "in 100,000 years elephants can go to the size of cows or sheep", says Helgen. And contrary to the oft depicted "March of Man" that portraying a one-way trip to increasing modernity, "evolution can go backwards", says geneticist João Teixeira at the University of Adelaide.[118] Fruit flies for instance can throw up mutations that take them back to sporting four wings – a characteristic of their ancestors. So it's not impossible that perhaps a mere 100,000 years ago *Homo erectus* made their way to one of the islands and not only shrunk like stegodons but produced throwbacks to a more ancestral state.

But the thing about *Homo erectus,* you'll recall, is the extreme diversity of the specimens who've been lumped into this category. It includes the diminutive, small-brained folk found in Dmanisi, Georgia who were the first to venture out of Africa, 1.77 million years ago, as well as Java's 108,000-year-old Ngandong population with their modern-sized brains.[119] The technological capabilities attributed to *Homo erectus* are also startlingly diverse. While the Dmanisi folk were limited to rough axes, others groups are considered responsible for ornamental carvings on mussel shells and exquisite tear-drop-shaped Acheulian axes. There are also hints that some groups of *Homo erectus* tamed fire.[120]

The champions of multiregional evolution have been jumping up and down for decades, saying *Homo erectus* was modernising in China and Java via a process they believed had to do with interbreeding. Could *Homo erectus* have been interbreeding with Denisovans? And could some of the more recent skulls themselves *be* Denisovans?

Figure 7. Feast your eyes on the most enduring piece of technology ever invented. Acheulian hand axes, such as this exquisite 400,000-year-old specimen found in the Hoxne brick pit in Suffolk, were crafted for around 1.5 million years by species including *Homo erectus*. John Frere, who found this specimen, commissioned this illustration for his publication in the journal *Archaeologia* in 1800.

John Hawks, who labels himself a "modern multiregionalist", thinks so. "The question I'm asking is, why should we think that these fossils are *Homo erectus*?"[121] Many of the more modern-looking skulls previously placed in the *Homo erectus* box are being taken out for another look. These include "Dali", a 250,000-year-old skull with some modern features found in China's Shaanxi province as well as the owner of some massive 125,000-year-old teeth found in 2015 in Xujiayao in Hebei province, and Penghu 1, a large, chinless jaw with big teeth, dredged up in a fishing net 25 kilometres west of the Taiwan coast in 2015. Other contenders included a 130,000–160,000-year-old tooth unearthed in Cobra Cave in the remote Annamite mountains of northern Laos.[122]

Until recently, John Hawks had considered Java's big-brained, 108,000-year-old Ngandong specimen the leading candidate for a Denisovan. And then in 2021, along came Dragon Man. The gigantic skull had reportedly been hidden in a well in Harbin in northern China. The labourer who dug it from a riverbank in 1933 stowed there to keep it from the Japanese invaders who'd pressed him into building a bridge across the Songhua River. Decades later, the bridge builder revealed the whereabouts of the skull to his grandchildren, who donated it to science in 2018. It was not just enormous; it was stunningly intact, with an enormous molar protruding from its upper jaw. It also possessed a remarkable mix of ancient and modern features, including thick brow ridges and one huge molar but a large, modern-sized brain. "I think this is one of the most important finds of the past fifty years," said Stringer, who collaborated on the project. Palaeontologist Qiang Ji, of Hebei GEO University, named it *Homo longi*, which means dragon man – and, together with an international team, gave him an estimated age of at least 146,000 years.[123]

Many suspect they already know the identity of the gigantic face with the weird tooth. "I think it's the best skull of a Denisovan that we'll ever have," says Jean-Jacques Hublin. "I think it probably is a Denisovan," agrees Stringer. But as we've seen, arguing over jaws and skulls doesn't get us very far. What we need is the revelatory power of DNA – or protein. I suspect Dragon Man could deliver any day. Indeed, while I was revising the final draft of this chapter in April 2025, a report emerged in *Science* that Penghu 1– the dredged-up Taiwanese jaw – is a Denisovan (based on its protein), with an estimated age of more than 50,000 years.[124]

And of course, it's impossible not to try to close the loop. Remember the ice age Aboriginal skulls – Talgai Man, Cohuna Man and the Kow Swamp People? Modern Aboriginal people and people of Papua

New Guinea have some of the highest levels of Denisovan DNA in the world. Their ancestors might have had more. Rather than the "mark of Java", perhaps the archaic Australians' features reflected the "mark of Denisova". Even Stringer, who has long argued that the ice age Australian skulls were extra thick in keeping with their overall large body size, rather than having inherited their robustness from an archaic ancestor, is open to the possibility. "I'm the first to acknowledge we don't have a full explanation. Is [their large skull and tooth size caused by] their Denisovan DNA? It might have been higher in the past," he told me.

Ghost species

Across Africa, Europe and Asia, fossils aged between 300,000 and a million years old show a mosaic of ancestral human features. They are a mishmash of different traits, such as the presence or absence of chins, protruding or flat faces, and crania that are flattened or rounded. For archaeologist Eleanor Scerri at the Max Planck Institute in Germany, they can be explained by the interbreeding of *Homo* ancestors who had previously been separated by barriers, such as deserts, that became passable during climatic shifts.[125]

That's the narrative suggested by genetics, too. Since 2018, analyses of some African populations have suggested that their ancestors interbred with ancient *Homo* populations. Genetics can't tell us exactly who those *Homo* groups were, since we don't have genomes of *Homo erectus*, *Homo heidelbergensis* and the like. So, these unidentified groups are called "ghost" ancestors.[126]

In March 2025, however, a group of geneticists from Cambridge, armed with a new algorithm, dug into the genomes of populations from around the world. They found that regardless of which population they looked at, each showed evidence of descent from two ancient ancestral populations. Those two populations had been separated for a million years, perhaps by glaciers or deserts.[127] Then, about 300,000 years ago, they found each other – and although they were different, they were not so different that they couldn't interbreed.[128] We are the result of that mixing.

Summing up

What do we make of all this? This chapter has traced the theory of human origins since Darwin's time through many of its twists and turns. It's made for a wild ride, especially over the last decade, and the terrain is still erupting. But there is a track.

It begins with the central enigma. One hundred thousand years ago, the *Homo* family was incredibly diverse. Now we are only one, *Homo sapiens*. Two main theories have vied to explain that enigma. One is that we are a species, descended from one common ancestor. The other is that our monotony derives from being so very well blended.

Until the 1980s, this was an unwinnable debate, fought on the evidence of fossils and stone tools, which respectively gave clues to the physical and mental attributes of our ancestors. Both were incomplete sources of evidence. The preservation of fossils is a chance event and can't possibly represent the true diversity of populations. Stone tools can't definitively identify new types of humans, either – technologies can be copied.

In 1987, mitochondrial DNA entered the fray, followed in 2000 by the full genome code. It didn't end debate, but for the first time it provided big data about ancient human populations. Instead of digging up the odd skull, the new breed of computational geneticists are doing archaeology by digging out fossil DNA from human genomes and cave sediments. Instead of arguing about the shapes of skulls, they are using the language of physicists, making testable predictions and designating confidence intervals around them.[129] Svante Pääbo, for instance, reflected on a conversation with David Reich in the late 1990s, when he estimated, there was a 10 percent chance that Neanderthals had contributed more than 5 percent to the modern human genome. "If we had been discussing the form, shapes, holes and ridges of Neanderthal bones, we couldn't have made any realistic estimate of how sure we were of what we had found. Neither could we have been confident of being able to collect more data to resolve the issue with greater confidence. With DNA we could."[130]

As Hawks put it, "Genetics has given us a yardstick. It's a fundamental problem with anthropology: we only have the tools we have. For the most part they work, when you have a lot of variation, say with molluscs or carnivores. But we don't have that with fossils. Getting DNA helps because we get variation back. That's a source we didn't have till now."

So where does that leave the two competing theories of human origins?

They're both right.

"Out of Africa" got it right in that the dominant input is from Africa, dating to a root that took hold around 300,000 years ago. But Darwin's tree structure, with species branching off never to meet again, is an oversimplification. Weidenreich's trellis, although incorrect in its details, is a more apt description of our evolution. With the discovery of more and more "missing links" that do not fit neatly onto a single linear branch, the idea of multiregionalism – local interbreeding to create hybrids – is being voiced more assertively. Particularly to explain the not-quite-modern human skulls that have surfaced across China in the past few decades.

"It is quite possible that this population represents gene flow between *Asian H. erectus*, and possibly *H. antecessor*, *H. bodoensis*, and/or early Neanderthals, supporting the idea of continuity with hybridization as a major force shaping human evolution in eastern Asia during the late Middle and early Late Pleistocene," wrote anthropologists Xiujie Wu from the Chinese Academy of Sciences in Beijing and Christopher Bae from the University of Hawai'i.[131]

It makes evolutionary sense. Footloose and fancy-free, splinter groups roaming the earth evolve rapidly because two powerful evolutionary pressures are operating at maximum force: the selection pressure to adapt to new environments, and the fact that new mutations are more likely to take hold in small populations.

So, some Neanderthals evolved fair hair and skin to better adapt to low levels of UV light, and a vamped-up immune system to protect them from transmissible diseases as they huddled up in their caves. Some Denisovans adapted to high altitude. When African bands came along and mixed with the Denisovans, they took an evolutionary shortcut to acquire these useful genes.

Many bands of humans have gone wandering out of Africa; most went extinct.[132] (See box: "Out of Africa, again and again".) But only one group, who left around 60,000–70,000 years ago, seems to have had what it took to thrive across the farthest reaches of the planet. That's the group that mixed with Neanderthals, Denisovans and, as new evidence suggests, other archaic ghost species.[133] (See box: "Ghost species.")

Mixing is how we became such a homogenous species – and the key to our success. Australia's Indigenous peoples, the first modern humans to trek out of Africa and survive at the ends of the earth, are living proof.

Out of Africa, again and again

"Out of Africa" conjures up the notion of a singular event – an intrepid tribe of savvy *Homo sapiens* marching out to conquer the world. Not so. Various *Homo* species have been marching out of Africa for millions of years. Like all hunter-gatherers, they needed to be on the move.

The first of them left more than 1.95 million years ago, perhaps leaving evidence of their activity in "cut marks" on fossilised animal bones in Romania. When it comes to modern humans – *Homo sapiens* – the date of their exodus poses a conflict: geneticists and archaeologists give us different dates. Geneticists tells us that all living people outside Africa arose from a relatively small population that left Africa between 50–60,000 years ago and soon interbred with Neanderthals. The suggested date of the interbreeding is 45,000–50,000 years ago.[134]

But archaeologists have identified the bones and artefacts of thoroughly modern humans outside Africa that are much older. The Israeli caves known as Qafzeh and Skhul are home to skeletons older than 90,000 years, while an upper jawbone at Misliya Cave dates to about 180,000 years. Greece wins the antiquity stakes with a 210,000-year-old skull found in Apidima Cave. Chinese fossils of modern human teeth found in Fuyan Cave in Hunan province date to older than 80,000 years, teeth in Lida Ajer Cave in Sumatra are around 70,000 years old and Tam Pà Ling Cave in Laos has head and jaw fossils which range in age from 86,000 years to 68,000 years old. In Australia, Madjedbebe rock shelter in the Northern Territory records sophisticated artefacts such as a ground-edged axe dated at 65,000 years. These people would have had to cross open seas to reach Australia.

One way to reconcile the archaeology with the genetics is that while there were multiple expeditions out of Africa, not all these bands of explorers were large enough or lucky enough to leave descendants. All modern humans outside of Africa appear to be the offspring of a group that left around 50–60,000 years ago. A 55,000-year-old skull in Israel's Manot Cave may represent the remains of a member of this successful group.[135]

Figure 8. An edited version of an iconic illustration published in 1965 by Time-Life Books. It suggested a linear progression in which ancestors went extinct, leaving the stage to the next *Homo* species. Now we know that many *Homo* species existed at the same time, and some interbred.

3

The Science of Medicine

It gradually dawned on me there that some things I had been taught in medical school were probably lethally wrong.

Iain Chalmers[1]

Medicine is a science of uncertainty and an art of probability.

Sir William Osler[2]

Historically, medicine has had a tenuous relationship to science. While attempts to apply the scientific method to medicine go back to James Lind's scurvy trials in 1740, rules for battling bias only appeared in the 1970s, with Scottish physician Archie Cochrane and the birth of evidence-based medicine. There is still a way to go, because two features of medicine resist the empirical incursion. One is the notion of the "art of medicine", which at its worst can give doctors a god complex. The other is the devious forms of bias that beset experiments conducted on humans.

Both factors are on full display in this chapter, which tells the story of how a cadre of doctors around the world promoted the worming medication ivermectin to treat Covid-19 as an alternative to vaccination.

...

"You are not a horse. You are not a cow. Seriously, y'all. Stop it."[3]

Street talk, tweeted from the normally staid communications office of the United States Food and Drug Administration, created quite a stir back in October 2021. Clearly it was a message trying to cut through. Indeed, it had to pierce a molasses of misinformation that claimed ivermectin, a worming treatment, was a cure for Covid-19.

By the time you read these words, I suspect the idea of using ivermectin for Covid will have gone the way of laetrile, the apricot kernel cancer cure, that was the craze of the 1970s. It turned out laetrile demonstrated no anti-cancer effect and was toxic. It has been described as "the most thoroughly studied failure in the history of medicine".[4]

Whatever lessons were learned, they did not hold back the ivermectin craze that blazed during the Covid pandemic.

I experienced it in a personal way when I discovered that several of my not-so-young friends were placing their hopes in a worming medication rather than a vaccine. I've been a science communicator for many years, but until that moment, communicating science had never been a matter of life and death. So, I set aside the book I was writing – about how theories are tested in science – and spent months digging up, and laying at the feet of my friends, the body of evidence showing why vaccines were a proven defence against Covid and why ivermectin was certainly not.[5] I counted some successes.

After months going off-piste, I realised I had the makings for another chapter of my book. Because what could be more important than proving whether a treatment works when a deadly virus is raging all around you? It turns out, this case study of ivermectin shines a spotlight on a bigger story: that the transformation of medicine into a science is far from complete.

Some readers may have come across the term "evidence-based medicine" and been taken aback. Isn't all medicine evidence-based?

Surely we're a long way from the era of bloodletting?

We are. Bloodletting was based on Hippocrates' theory of the four humours – the blood, black bile, yellow bile and phlegm – whose proportions were balanced by bleeding the patient. From 500 BCE until as recently as the 1920s, it was used to treat everything from fever to high blood pressure. One of today's leading medical journals, *The Lancet,* took its name from the fine-tipped blade that was the physician's stalwart.[6] Mercifully, today's medical theories are based on sounder scientific principles.

But there are theories, and then there is evidence.

In the seventeenth century, the scientific revolution dethroned theories and placed evidence in the seat of power. When Newton developed his laws of gravity to describe the measurements Kepler had made, he famously wrote: "*hypotheses non fingo*" – "I frame no hypotheses".

In medicine, this revolution did not take place until the mid-twentieth century – a lag of 300 years! The case of ivermectin shows that the revolution is not over.

In a Petri dish down under ...

Melbourne's Peter Doherty Institute came into being in 2014 to provide counterintelligence against a global threat: viruses, the most numerous and devious biological entities on the planet. The strategy, dreamed up by Melbourne University Deputy Vice-Chancellor Jim McCluskey and Royal Melbourne Hospital virologist Mike Catton, was to combine the forces of their academic and clinical institutions.

Nestled amongst Melbourne's elite medical institutes, the twelve-storey institute exemplifies the exuberant architecture metastasising across the Parkville precinct. From its climate-proof glass and timber-slat facade to the innermost high-security lab where scientists in spacesuits work with deadly viruses, every aspect exudes a sense of mission.

Melburnians will recall that the prominent site was once home to Ampol House, its iconic eight-floor spiral staircase visible through a curving glass wall. Some protested the demolition. McCluskey countered that public good would be better served by research than heritage. The building would be "a medical centre of global importance and would tackle threats such as influenza pandemics".[7]

In 2020, the Doherty Institute fulfilled this prophecy, although the threat wasn't the dreaded influenza. In mid-January of that year, Sharon Lewin, the institute's congenial director, was spending her annual break hiking the gruelling W Trek that winds between the massifs of Torres del Paine in the wilds of Patagonia. With internet connections sparse, it was up to her deputy director, Mike Catton, to respond to Australia's first case of Covid.

On 19 January 2020, a 58-year-old man arrived in Melbourne from Wuhan, feeling unwell. On 24 January, the man was admitted to Monash Medical Centre with fever, cough and worsening shortness of breath. Catton's team analysed a nasal swab using a PCR test and found that it carried a virus that matched the new Wuhan strain.

Within five days of the man's admission, the team achieved two things. They grew the virus in a dish of African green monkey kidney cells, and they read its genome. Using the nasal sample, researcher Leon Caly infected a dish of monkey cells with the virus. As the days went by, the monkey cells, growing like a lawn on a Petri dish, grew ragged. When the fluid above the lawn was examined under an electron microscope, tiny spheres bearing crowns of spikes were visible – coronaviruses. (The Latin word for crown is *corona*.) This coronavirus soup allowed Caly to isolate the virus genome and read every letter of its 29,811-letter code.

The team's achievement allowed two important things to happen.

First, this genome from Patient Zero would allow researchers to trace the spread of the virus in Australia. While the original sequence had been broadcast on 10 January 2020 by Eddie Holmes at the University of Sydney and Zhang Yongzhen at Fudan University, the virus accrues mutations as it spreads.[8] Patient Zero's genome would act as the baseline to trace chains of transmission, akin to the way linguists trace the spread of languages by studying letter changes in words. For instance, the spread of European languages can be tracked through changes in the word for water: wadr, wasr, water.

The second important thing was that at that moment, outside of China, they were the only lab in the world that had succeeded in growing the virus. Being able to grow the virus was the first step towards finding its Achilles heel and developing drugs and vaccines to target it.[9]

Catton alerted colleagues around the world of his team's success on 29 January, and within twenty-four hours requests came pouring in to share their viral soup. Months later, Catton, Julian Druce and Caly would win a national prize for their work.[10] But for Caly, the limelight would soon take on a much more glaring hue.

Enter ivermectin

Twenty kilometres southeast of the Peter Doherty Institute lies Monash University and the lab of Kylie Wagstaff, who is married to Caly. She was also keenly interested in the new coronavirus.

Wagstaff is a rising star with a string of awards under her belt. Her eighteen-year career has focused on how information travels between a cell's nucleus, where its genome is vaulted away, and the surrounding cytoplasm, where strings of code instruct tiny protein-based machines in the day-to-day operations of the cell. When a virus invades a host cell, such as one of our lung cells, it uses its own code to reprogram the host machinery to make copies of the virus. Wagstaff's strategy is to track down the compromised host machines and throw a chemical spanner into their works. Back in 2011, she and colleague David Jans discovered that one such chemical spanner was a drug called ivermectin.

Until the Covid-19 pandemic, most people had never heard of ivermectin. Which is somewhat surprising, since its discoverers, Japanese microbiologist Satoshi Ōmura and American parasitologist William Campbell, were jointly awarded the 2015 Nobel Prize for their efforts. Ivermectin was the world's first effective treatment for a horrific set of parasitic diseases: onchocerciasis and lymphatic filariasis. Their common names, river blindness and elephantiasis, give a mere hint of the suffering they cause. River blindness also causes a maddening itch, while elephantiasis is named for the elephantine swelling of not only the legs but also the genitals. Endemic to tropical regions in Sub-Saharan Africa and South America, the diseases are caused by microscopic nematode worms that are transmitted by the bites of black flies in the case of river blindness and mosquitos in the case of elephantiasis.

To find a treatment for these diseases, Ōmura turned to a time-honoured source: the soil, where microorganisms of all stripes battle each other with chemical weapons. He tested thousands of strains of bacteria and found that one, *Streptomyces avermitilis,* carried a weapon against parasitic worms. His colleague Campbell showed that it was safe and effective for treating parasitic worms in animals. Chemists purified the active agent and increased its potency to produce the wonder drug ivermectin. Taken once a year, it is extremely effective. As of 2016, river blindness was effectively eliminated in Colombia, Ecuador, Mexico and Guatemala thanks to the distribution of ivermectin via the Carter Center and partners.[11] The discoverers of ivermectin deserved every cent of their Nobel Prize money.

Wagstaff's team had shown that as well as vanquishing parasitic worms, ivermectin might also be enlisted to the cause of fighting viruses. For instance, it slowed the growth of dengue virus in a Petri dish. Importantly, however, clinical trials in Thailand showed that ivermectin did *not* knock down the level of dengue virus in the blood of sick people, nor did it speed their recovery.[12]

In February 2020, like every other scientist in the world, Wagstaff was eager to deploy her skills against SARS-CoV-2. Since Caly had already successfully grown the new coronavirus at the Doherty Institute, they decided to collaborate to see if ivermectin would halt its growth.

It did. Forty-eight hours after ivermectin was administered, the African green monkey kidney cells were largely rid of the virus. But there was a rub: the concentration of ivermectin required to kill the virus was over twenty times higher than what could be achieved in a patient. It's not just that such a dose would be toxic to people; it would be impossible to achieve. In the human body, ivermectin is sponged up by blood proteins, broken down and cleared.

"If Nobel Prizes were handed out for curing life-threatening diseases in a Petri dish, then I … and virtually every translational research scientist … would have one," noted Jorge Caballero, co-founder of Coders against Covid.[13]

Experienced drug developers such as Craig Rayner, formerly at Monash University and now with the US-based pharmaceutical consulting firm Certara, were highly sceptical that ivermectin had anything to offer as far as treating Covid.[14]

Nevertheless, a paper with Caly as first author was published online on 3 April 2020. It was titled "The FDA-approved Drug Ivermectin Inhibits the Replication of SARS-CoV-2 *in vitro*" (*in vitro* means "in glass" in Latin and refers to the Petri dish).

In the fast-streaming channels of the internet, it set off a global shockwave. Its biggest impact was felt in Peru.[15]

Peru's perfect storm

A month after Caly and Wagstaff's paper appeared, Peru became the first country in the world to include ivermectin in its national guidelines for the treatment of Covid-19.

Did it help?

If you google which country had the highest per capita death rate from Covid, the answer is … Peru.[16]

Its infection rate was similar to that of many other countries. So too were the problems of a crowded population, multigenerational households, a weak healthcare system and insufficient bottled oxygen to go around. But Peru's death rate of 646 per 100,000 people was more than twice that of Brazil and the USA.

Patricia Garcia, a former health minister of Peru, suspects the uncontrolled use of ivermectin, along with hydroxychloroquine and steroids, is a big part of the explanation.[17] "Unfortunately, we practise polypharmacy. All these medications were available over the counter in every single pharmacy – for cents." Ivermectin, in particular, was familiar to people. They were already using it once or twice per year to treat parasitic infections, she explained. Based on this regular use, health officials considered it safe, and believed it was better to do something than nothing. The country didn't have time "to wait for scientific evidence" said Víctor Zamora, Peru's health minister. "What do you do? Give them water?" remarked Eduardo Gotuzzo, the former director of the Alexander von Humboldt Institute of Tropical Medicine in Lima.[18]

Garcia shared their sense of desperation but channelled it into clinical trials. "My colleagues did not understand the importance of 'evidence-based medicine.' At the university, I teach my students the harms of just-in-case medicine, and that anything you do has an effect." Surveying the wreckage of Peru's pandemic statistics, she is determined to capture the lessons. "This should never happen again," she tells me.

Eminence-based medicine

Garcia was happy to indulge me in an hour-long Zoom interview in late 2021.[19]

It is a sad irony that her career has been devoted to rolling out evidence-based medicine to benefit Peruvians – but when the pandemic struck, she was overpowered by a tide of what she refers to as "eminence-based medicine".

Garcia wanted to be a doctor from the age of five, partly inspired by her Andean grandmother, who taught her traditional medicine, and partly by the months she spent in hospital with a severe brucellosis infection from eating non-pasteurised dairy products.[20] The first of her family to go to university, she earned her medical degree from Lima's Cayetano Heredia University, then did her residency at the University of Washington in Seattle. She was hoping to secure lung cancer treatment there for her father, a journalist. Sadly, he died before that was possible. But in Seattle she met a mentor, epidemiologist King Holmes. He focused on sexually transmitted diseases, particularly HIV. "One of the things he told me is that public health is important. Because with one action, you can change the life of thousands of people."

Garcia switched course. Her early work involved taking boats deep into the Peruvian jungle to survey women's sexual health. Whether they lived in the jungle or in cities, Garcia found they relied on local pharmacies to treat sexually transmitted diseases. She returned to Lima in 1997 with a mission to improve sexual health, particularly by improving the skills of pharmacists. And so began her meteoric rise.[21] From working for a government agency, to heading the epidemiology unit at Cayetano Heredia University, then heading the Peruvian National Institute of Health. As Minister of Health from July 2016 to September 2017, she rolled out HPV vaccination, electronic medical records and telemedicine. "I had the opportunity to do something good," she told me.

In early January 2020, with Covid looming, Garcia, now back at Cayetano Heredia University, helped the government prepare. They established two task forces, one focused on clinical treatments and one, headed by Garcia, on new techniques. "There were so many crazy proposals, like using a magnet to make a diagnosis." By the middle of May, 500 people were dying each day. Many of them were doctors; their colleagues displayed their pictures outside the medical college in Lima.

Peru enforced lockdowns, but they couldn't last for more than a couple of weeks. With 70 percent of the population employed as street vendors or drivers, and many households without refrigerators, putting food on the table meant being out and about.

Doctors turned to two glimmers of hope. One was a paper from Marseilles published on 17 March 2020. It described how the malaria medication hydroxychloroquine, when given to twenty seriously ill patients, reduced their levels of virus. The other, published on 3 April, was the Australian ivermectin paper by Caly and Wagstaff.[22]

The title of the Australian paper was riveting: "The FDA-Approved Drug Ivermectin Inhibits the Replication of SARS-CoV-2 *in vitro*". Many took that to mean the FDA had approved the drug for Covid. "I had to explain to people: it's only FDA-approved as a worming drug, not for Covid," recalled Garcia. When Garcia dug into the paper, she discovered another major caveat: the enormous dose required. Would these results mean anything at all for Covid?[23]

The answer arrived three days later, on 16 April, by way of an online "preprint". It showed ivermectin was stunningly effective.[24]

Preprints

Let's take a moment to talk about preprints. Preprints are first drafts of scientific papers that are aired on "preprint servers" while they await peer review. Peer review is analogous to having building inspectors certify the integrity of a new home. Two or more experts read the paper, looking for weaknesses in the methods or the conclusions drawn. Without peer review, it's hard for non-experts to assess the integrity of the research.

Preprint servers publish papers before they have been through this process. Some but not all preprint servers carry a warning that the paper has not yet been peer reviewed. For instance, medRxiv, a server for medical preprints, carries this warning:

> Caution: Preprints are preliminary reports of work that have not been certified by peer review. They should not be relied on to guide clinical practice or health-related behavior and should not be reported in news media as established information.

So why publish preprints at all? The process of peer review takes months, and many journals restrict access to published papers behind a paywall. During a pandemic, there's a strong argument for getting life-saving information out fast and freely – there are no paywalls on preprint servers. For instance, in early March 2020, researchers at Oxford University led a trial to test four "repurposed" drugs – drugs approved for other conditions – on gravely ill Covid patients in a lightning-fast trial known as RECOVERY.[25] They found that the steroid dexamethasone, which tamps down friendly fire from the immune system, reduced the death rate by about a third. Hydroxychloroquine, on the other hand, not only failed to save patients; it increased the death rate. These results were hastened onto the preprint server medRxiv in June and July, several months before the peer-reviewed papers appeared in *The New England Journal of Medicine*.[26] These rapid preprints, showing that dexamethasone worked and hydroxychloroquine didn't, saved lives. By one estimate, a million lives were saved in 2020 thanks to the quick verdict on dexamethasone.[27]

Too good to be true

Now let's go back to the ivermectin preprint that riveted Garcia to her computer screen on 16 April. Researchers usually take preprints with a grain of salt but the first months of the pandemic were desperate times. Around the world, doctors watched helplessly as patients asphyxiated from pneumonia or their bodies shut down, as one after another of their organs failed. Thousands were dying every day, mostly the elderly, but also many younger people. This preprint promised salvation. And on first glance, it smacked of credibility. The senior author, Mandeep Mehra, was an eminent cardiologist and director of the Brigham and Women's Hospital Heart and Vascular Center at Harvard University.

The research was also technologically dazzlingly, employing big data and artificial intelligence. It had harvested the medical records of 100,000 patients from 700 hospitals on six continents who'd been placed on ventilators. A Chicago-based company, Surgisphere, had used AI to compare the outcomes for those who were given a single standard dose ivermectin with those who weren't. Those who received the ivermectin tablet were *six times* less likely to die than those who had not.

"When I read Mehra's paper, several things didn't make sense to me," Garcia told me. She phoned a friend at the Harvard School of Public Health and asked him to connect her with Mandeep Mehra. She recalls him saying, "You're not the only one with concerns."

...

Venezuelan-born Carlos Chaccour had beaten her to it. An intense young public health physician, he's also an ivermectin expert. As you tend to be, if you're a doctor to jungle communities. Straight out of medical school at the Central University of Venezuela, he had flown, jeeped, motor-boated and paddled south to reach the otherworldly Venezuelan Amazon depicted in the movie *Up*, to tend to patients who lived in villages dotted amidst the dense jungle and towering mesas.

Three years later, the jungle-tested doctor enrolled in a master's degree at the London School of Hygiene and Tropical Medicine. He focused on malaria, a particular problem for modern-day Venezuela. Back in the 1960s, the country had beaten the disease through a decades-long mosquito-spraying campaign. But over the last decade, along with the economy, mosquito control had collapsed. Malaria returned with a vengeance, reaching case numbers eight times higher than that of neighbouring Brazil, mostly south of the Orinoco River, where communities live with very little in the way of health care.[28]

Unfortunately, these communities can no longer rely only on bed nets because the local mosquitos have evolved a new strategy. Instead of hiding in homes during the day to feast on the sleeping inhabitants at night, they are biting people outdoors as they go about their business at dawn and dusk. An alternative solution is needed. Chaccour thought of ivermectin. In South America it is often used to treat people with scabies, an itchy skin condition caused by a microscopic burrowing mite. When a person takes an ivermectin pill, their blood poisons the mite. Chaccour wondered if a similar strategy might be useful for poisoning mosquitos. To find out, he organised a trial. Human subjects volunteered to take an ivermectin pill and then place their arm in a cage full of mosquitos. The mosquitos were then tracked, to see if they died after feeding on ivermectin-laced blood. They did.[29] One of the intrepid volunteers was Chaccour's future wife.

By the time the Covid pandemic arrived, Chaccour had a great deal of experience with ivermectin. So when the Australian paper by Caly and Wagstaff appeared, Chaccour, now based at the Barcelona Institute for Global Health, scrutinised it intensely. Five days later, he and three international colleagues sent an article to *The American Journal of Tropical Medicine and Hygiene* titled: "Ivermectin and Covid-19: Keeping Rigor in Times of Urgency".[30] It argued that while ivermectin had a good safety record, there were reasons to be cautious. Extremely high doses were toxic to the brains of mites and mosquitos. They might also be toxic to the human brain, especially to patients with Covid, which can cause blood vessels in the brain to become leaky. Careful clinical trials were needed, Chaccour and his colleagues concluded.

Surgisphere oddities

They had barely finished typing their article when the results of just such a clinical trial appeared in the form of the preprint from Surgisphere. Chaccour was perplexed. The preprint described seriously ill Covid patients who'd been treated with ivermectin in hospitals around the world between 1 January and 1 March 2020. But in that time period, ivermectin as a treatment for Covid was not on the radar. It was not till the Australian paper had appeared, on 3 April, that the medical Twitter streams and chat groups Chaccour frequented made any mention of ivermectin. It was just the first of many oddities.[31]

Surgisphere's African database reported that three patients had been hospitalised, put on ventilators and given ivermectin. But between January and March, official records showed that across the whole African continent, only two Covid cases were reported. Chaccour, who is familiar with African hospitals through his experience testing ivermectin on malaria, also found it beyond belief that so many of them were equipped with electronic systems that instantly relayed their records to a US database.

His scepticism extended beyond Africa. How had Surgisphere, a tiny and previously unknown company based in Chicago, been able to access 100,000 patient records from 700 hospitals on six continents so seamlessly? Procuring patient data, even if it's de-identified, is a fraught process. Different countries impose different rules. The preprint failed to give details of which hospitals had partnered with Surgisphere. In Australia, infectious disease experts such as Allen Cheng, based at the Alfred Hospital, had never heard of the company, nor had anyone from his hospital, one of Melbourne's major emergency centres.[32] The prestigious US-based Stanford University, which Surgisphere claimed as a partner, was similarly ignorant of the company. Moreover, in Australia as in Africa, the Surgisphere data didn't fit the numbers on the ground. In Victoria and New South Wales – the states most affected by Covid – the Surgisphere numbers disagreed with the health departments' records of hospital admissions and deaths.[33]

For Chaccour, the reddest of the red flags was the size of the reported effect of ivermectin. Untreated patients had a death rate of 8.5 percent; in those treated with ivermectin, according to Surgisphere's results, it was down to 1.4 percent – a sixfold decrease. "I've been working with ivermectin for twelve years," Chaccour told *The Scientist*. "This doesn't make any sense ... this massive effect."[34] Chaccour finally received a response to his queries from the founder of Surgisphere, Sapan Desai. "Let's just say my concerns about the study did not reduce whatsoever."[35]

Physician's Covid combo

Chaccour and Garcia shared their concerns with each other, and Garcia shared them with her colleagues on the clinical task force in Peru. Some, especially the younger doctors, agreed there was insufficient evidence to recommend ivermectin. But she was unable to hold back the majority vote. On their advice, the Peruvian government placed ivermectin and hydroxychloroquine on the national guidelines in May 2020. Garcia and several other members of the clinical task force resigned in protest. Then they watched helplessly as ivermectin frenzy took hold.

Gustavo Aguirre Chang, a local doctor, posted glowing endorsements on a website hosted by an ivermectin manufacturer. A "physician's Covid combo" became popular: ivermectin, azithromycin, prednisolone, hydroxychloroquine. On TV, radio and social media, influencers began prescribing their own regimes – for instance, recommending people take ivermectin *three times a day* for prevention. Recall the standard dose was once or twice per year.

The government sent the army to roll out ivermectin across the country. Evangelical groups sent volunteers into jungle communities armed with injections of ivermectin, claiming it was equivalent to a vaccine.[36] When pharmacies ran out of ivermectin, people turned to veterinary formulations. Hospitals began admitting people with stomach and liver problems, tinnitus and dizziness – all cases of ivermectin poisoning. Worse were the pictures Garcia received of Indigenous people from jungle areas with necrotic ulcers on their arms. "They had been injected with veterinary-grade ivermectin. It was one of the worst things I've ever seen."

For most ivermectin users, the harm was more insidious. Believing themselves to be protected, they went about their daily lives on the streets. They were given "a false sense of security", Garcia told me.

Ivermectin fever wanes in South America …

By the time Chaccour and Garcia connected, they were lone holdouts in a world polarised by ivermectin. In one camp, clinicians championed its use. In another, major funders such as the Gates Foundation and WHO deemed trials pointless, since the Australian study showed the ivermectin concentrations necessary to kill the virus were so high, they could never be achieved in people.[37]

Chaccour and Garcia disagreed with both camps. In May 2020, Chaccour launched a trial at his institute in Barcelona. Garcia started another in Peru. Both would be randomised controlled trials (RCT), where the patients are randomly assigned to either a treated or an untreated "control" group. Chaccour and Garcia found it almost impossible to recruit patients – but for very different reasons.

Chaccour had started planning a trial in early April, when Spain had around 8000 cases and 800 deaths each day, a surge that overwhelmed Madrid's morgue and required the ice rink to hold the dead. But within weeks, the deadly wave receded. By July, there were close to zero cases in Navarre, the province where the trial was planned. "It was impossible to recruit," Chaccour told me.[38] Meanwhile, online trolls were accusing him of holding back the positive data. "You're an expert; you know it works," said one. Chaccour, they claimed, was part of a "big pharma" conspiracy to suppress ivermectin in favour of more profitable vaccines or drugs yet to be developed.

Garcia had a different problem. To carry out her Peruvian trial, she needed a control group of people who had not used ivermectin in the past few months. That proved impossible. By one estimate, 70 percent of the Peruvian population were using it.[39] She had to compromise and enrol people who had not used ivermectin during the previous month. Meanwhile, she watched in dismay as other countries followed Peru's example: Bolivia, Paraguay, parts of Brazil, Slovakia, South Africa, Zimbabwe.[40]

Within months, the flimsy evidence base supporting hydroxychloroquine and ivermectin began to topple. A WHO trial for hydroxychloroquine (Garcia was the lead investigator for Peru), showed that the drug did not reduce the death rate, while the Oxford-based RECOVERY trial mentioned above showed it slightly increased the death rate. The data from Surgisphere went up in smoke. Given the intense questioning of their findings, Mandeep Mehra, the Harvard cardiologist, and his co-authors asked Surgisphere for the raw hospital data on which the studies were based. When Surgisphere declined, the co-authors requested the withdrawal of their ivermectin preprint. The preprint, however, was just the tip of the iceberg. Surgisphere's dodgy database had been used for two other studies that were published in two of the world's most elite medical journals, *The Lancet* and *The New England Journal of Medicine*. Both papers were retracted.[41] In October, Peru removed ivermectin and hydroxychloroquine as treatments for severely ill hospitalised patients from the national guidelines.[42] By February, they were removed as a recommendation for treating anyone with Covid.

Chaccour would eventually publish a randomised control trial of twenty-four patients in January 2021.[43] It was inconclusive. It did not find any difference in the patients' viral load after seven days of treatment with ivermectin. However, he did find a small, non-statistically significant result. The treated patients seemed to recover slightly more quickly, and fewer of them lost their sense of smell, a result also seen in hamsters treated with ivermectin. "These results pointed to the need for additional data," observed Chaccour.

Garcia's trial found that ivermectin had no effect.[44]

... And blooms in the West

But as hydroxychloroquine and ivermectin use faded in Peru, ivermectin fever bloomed anew in the wealthy West, even as its citizens were offered some of the most effective vaccines ever to be developed. (Vaccines are traditionally considered effective if they offer 50 percent protection. The AstraZenica, Pfizer and Moderna vaccines were 80 percent, 90 percent and 95 percent effective.)

In September 2021, when the US FDA issued its "you are not a horse" tweet, every American had access to a free shot of Pfizer or Moderna. Given Peru's experience, Garcia found it "crazy" that other countries were still rushing to embrace ivermectin. Yet the Western frenzy was not fanned by a group we generally consider crazy. Those fanning the flames were a group of "eminent" doctors.[45]

There is much to analyse about the ivermectin moment. For one thing, the researchers who gathered evidence to see if ivermectin actually benefited sick people found themselves, for the first time in their lives, the targets of vitriol.

"This is not about science but psychology," Ed Mills told me. An epidemiologist at McMaster University in Canada, he led the TOGETHER trial, the first of the large and rigorous studies to test whether ivermectin could prevent severe illness. It found ivermectin was ineffective.[46] For their efforts, Mills and another member of his team, Craig Rayner from the US-based pharmaceutical consulting firm Certa Inc., were rewarded with hate mail. "It's not unusual to see a death threat in my inbox [since the trial]," Rayner told me. "It's pretty traumatising. We're all feeling that way."[47]

The volcanic heat around ivermectin was generated by outrage that a cheap, off-patent "cure" was allegedly being suppressed to benefit drug companies, who were developing far more lucrative Covid treatments and vaccines. Such a conspiracy required a cast of thousands – government health authorities across the world, the WHO and the overwhelming majority of the world's leading infectious disease and drug development experts – all colluding to supress the truth. The irony was that testing cheap, off-patent drugs was the chief focus of several large government, private and philanthropically funded trials at that moment.[48] Besides testing ivermectin, the TOGETHER trial was testing hydroxychloroquine, the diabetes drug metformin and HIV drugs, all of which proved ineffective. The antidepressant fluvoxamine cut hospitalisations by 30 percent, but this finding was unfortunately not replicated by other studies.[49] One of the high-profile conspiracy theorists was Bret Weinstein, an evolutionary biologist and former college professor. He aired his views on a now infamous episode of *The Joe Rogan Experience*. "If the WHO or CDC [Centers for Disease Control and Prevention] were to be captured; if influence was to be exerted over one or both of these bodies, surely we would need to talk about that ... that an essential set of organisations that are supposed to be protecting public heath might be doing someone else's bidding instead," he said.[50]

It is unsettling to hear a scientist subscribing to deep-state conspiracy theories. Eminent doctors also got on board. Joining Weinstein on Rogan's podcast was Pierre Kory, a lung and ICU specialist. Kory became one of the most prominent American flagbearers for ivermectin.[51] In December 2020, he proclaimed ivermectin a "miracle drug" against Covid-19 during a testimony to the US Congress that attracted over a million viewers online.[52] He is president and co-founder of a group of eminent specialist physicians – among them pulmonary and critical care specialists, kidney specialists and endocrinologists – known as the Front Line Covid-19 Critical Care Alliance (FLCCC). Kory's embrace of ivermectin cost him two American hospital positions in the space of a year.[53]

Ivermectin advocacy was not restricted to American doctors. The UK had Tess Lawrie, a doctor and prolific online influencer who founded the British Ivermectin Recommendation Development (BIRD) Group. In Melbourne, a small group of doctors gathered under the banner of the COVID Medical Network. In Sydney, gastroenterologist Thomas Borody, whose website proclaims him the inventor of modern treatments for ulcers, spruiked his ivermectin cure for Covid-19.

When it came to vaccines, these doctors expressed great concerns about possible side effects. But their medical reserve was nowhere to be seen when it came to the curative powers of ivermectin.

An Ohio-based lung specialist, Fred Wagshul, a founding member of FLCCC, claimed that the science behind ivermectin's use in Covid-19 patients was "irrefutable". He took part in legal action to force a hospital to administer ivermectin to a sick patient against the recommendations of the FDA and CDC. Wagshul told an Ohio newspaper that the CDC and FDA were engaged in a conspiracy to block ivermectin use in favour of Covid-19 vaccines and that the Covid death toll was higher as a result: "If we were a country looking at another country allowing those [Covid-19] deaths daily … we would have been screaming, 'Genocide!'"[54]

For many layfolk, the eminence of Wagshul and like-minded colleagues was enough to convince them that ivermectin must really be the cure they said it was. Yet Wagshul's "irrefutable" evidence is quite straightforward to refute. Perhaps they didn't teach evidence-based medicine at Wagshul's medical school. That wouldn't surprise Paul Glasziou, director of the Institute for Evidence-Based Healthcare at Bond University in Queensland, who previously headed Oxford's Centre for Evidence-Based Medicine. "Globally, evidence-based medicine is not well taught in many medical schools," he told me.[55]

As of January 2022, there were some eighty-six clinical trials investigating whether ivermectin is helpful or harmful for people with Covid-19.[56] They didn't agree. That's not unusual. This is where evidence-based medicine comes in. It was developed to sort through the mishmash of conflicting medical evidence. It does this by subjecting studies to quality control such as checking them for potential bias and whether the studies have the statistical power to support their claims. Had this quality control been applied to ivermectin, the harm done in Peru and elsewhere might have been avoided. To understand why, let's leave ivermectin for a moment, and take a brief tour of the history and principles of evidence-based medicine.

What is evidence-based medicine?

There are various wordy definitions. Let's go with "using the scientific method to measure the effectiveness of medicines".

Despite the efforts of scientific doctors such as James Lind, whose scurvy trial identified the benefits of lemons in the eighteenth century, and Alexander Hamilton, who identified the harms of bloodletting in the nineteenth century, it was only in the mid-twentieth century that the movement gained steam, with the first randomised control trial or RCT to test the effectiveness of streptomycin for treating tuberculosis, published in 1948. An RCT randomly assigns patients to two groups, one of which will receive the treatment being tested and one of which will not. The term "evidence-based medicine" was only coined in 1991 by Gordon Guyatt at McMaster University, as he sought a title for the course he was teaching to medical residents. His first choice, "scientific medicine", miffed his fellow doctors because it implied that their current practice was less than scientific.[57]

To be sure, their practice was informed by scientific theories. By the early twentieth century, medicine had finally done away with practices like bloodletting and the antiquated theories it was based on, and welcomed instead scientifically based theories such as Pasteur's germ theory, Jenner's theory of vaccination and Starling and Bayliss's hormone theory. By the 1950s, these theories delivered treatments that indisputably saved millions of lives. A sulpha drug, an early antibioticmade from coal-tar dyes, cured Winston Churchill's pneumonia, and penicillin (made from mould) treated the syphilis of WWII soldiers, perhaps helping the Allies win the war, as they mass-produced penicillin in a second "Manhattan Project". Smallpox vaccines saved an estimated five million people each year in epidemics worldwide. Childhood diabetes stopped being a death sentence thanks to the hormone insulin, and cretinism – a stunting of mind and body – became treatable because of the availability of the thyroid hormone thyroxine.

So, twentieth-century medicine was being informed by scientific theories. But the practice of medicine itself was not *scientific*. There was no requirement to test whether a treatment did more good than harm, even if that treatment was based on sound theories. "We had *mechanism-based medicine*, but the need to test medicines empirically was not the rule and still isn't, says Paul Glasziou. It was the likes of Newton and Galileo, with their careful measurements and experiments, who gave birth to science as we know it. Since then, to quote Michael Strevens, a professor of philosophy at New York University, "the iron rule of science" has been empiricism, the testing of theories with evidence.

For medicine, it seems to have taken another 300 years for there to be an equivalent rule. It was only in 1962 that the US FDA made it a legal requirement that newly developed drugs show proof of efficacy[58], and other countries followed suit. Even after that FDA ruling, many accepted treatments turned out to cause more harm than good. For instance, in the 1980s, if you had a heart attack, the ambulance medics would likely have given you a shot of lidocaine, a so-called antiarrhythmic, to restore the steady rhythm of your heart. While some radiologists had seen clots in the arteries of heart-attack patients, the blockages were considered a consequence, rather than the cause, of the heart attack. The reigning medical wisdom was: treat the arrythmia. But in 1988, a famous randomised control trial showed that if a catheter loaded with the "clot buster" streptokinase was snaked into the artery within hours of a heart attack, it reduced deaths by one quarter. Thinning the blood with oral aspirin had about the same size effect. The two together halved the risk of death.[59] As for lidocaine, a systematic review in 1999 showed that its use during a heart attack was more likely to kill you.[60] According to one estimate, up to 100,000 people died from this treatment.[61] The doctors who recommended lidocaine for heart attacks were acting on the basis of a good theory. The problem was, they hadn't tested it. "Mechanism-based medicine was killing people," says Glasziou.

Evidence-based medicine delivers "compound interest"

Evidence-based medicine is full of surprises. Besides revealing that some common treatments do more harm than good, it has also shown that most beneficial drugs have a small effect. But that's not as dismaying as it seems.

This finding emerged in 2013, when Glasziou and a group of colleagues surveyed the outcomes of hundreds of RCTs across the spectrum of medicine and published their findings in *Nature*.[62] They found that just over half of the medicines studied delivered a clinical benefit, while half performed worse than existing treatments. For those that did deliver a benefit, it was usually a small one.

Yet, like compound interest, the incremental gain delivered by each new treatment adds up. Glasziou points to the example of childhood leukemia. A death sentence in the 1950s, today the cure rate approaches 90 percent. But that accrual took five decades and many small discoveries, each delivering only a 2–5 percent improvement.[63] This huge advance over a period of fifty years was only possible because of evidence-based medicine. At every step, the effect of each new treatment was measured to test whether it did more harm than good. Empiricism provided the staircase for progress, as it had been doing for science since the seventeenth century.

Why did medicine resist the scientific method?

So why did medicine lag science by 300 years? It might have something to do with the "art of medicine". According to the *Cecil Textbook of Medicine*, medicine is:

> a profession that incorporates science and scientific methods with the art of being a physician. The art of tending to the sick is as old as humanity itself. Compared with its long and generally distinguished history of caring and comforting, the scientific basis of medicine is remarkably recent.[64]

Of course, when you're seated in front of your doctor, you want them to practise their art. After all, you're not a lab rat but an individual with your own quirks and volition. "Yes, there are guidelines, but I have to nuance them to my patients," says Allan Lew, a friend and cardiologist at Cedars-Sinai in Los Angeles, referring to how he decides whether to recommend aspirin for his individual heart patients. Nevertheless, science is encroaching on Lew's art: algorithms are now helping him to work out the risk–benefit ratio for individual patients, he tells me.[65]

Patients may also have "comorbidities", an assortment of conditions that means their needs do not necessarily match those of the patients in a clinical trial. Christine McDonald, a friend who is the director of the Department of Respiratory and Sleep Medicine at Austin Health in Melbourne, told me: "Besides lung disease, my patient may be experiencing partner abuse, the stress of being a single mother, or obesity ... They don't want to know about evidence-based medicine, they want the doctor to advise them."[66]

John Matthewson, a medical graduate who now teaches philosophy at Massey University in New Zealand, recalls an impressive demonstration of the art of medicine that he witnessed as a medical student. A patient walked in with a vague set of symptoms, but the doctor somehow realised that the problem lay hidden elsewhere. The doctor asked the patient how his toe felt. The surprised patient, who had not mentioned his toe, confessed that it hurt. Sure enough, when he took off his shoe and sock, the toe was blue. Poor circulation, just as intuited, although the doctor could not articulate exactly what had cued him.[67] But some doctors take their art to an extreme. "We didn't believe in a [randomised trial] ... If you've been doing this for decades, and you trust your assessment of the disease and your knowledge of medicine, it's OK to doctor," remarked Pierre Kory, America's chief ivermectin advocate, about another unproven treatment developed by his colleagues.[68] Kory seems to be harking back to the era of the paternalistic, all-knowing physician practising his art. To be sure, at their best an "artful" doctor is the kind who ferrets out a hidden condition based on intuition and has empathy with their patient. At worst, an all-knowing doctor risks developing a "god complex".

Archie Cochrane diagnoses the "god complex"

The "god complex" is a condition Archie Cochrane diagnosed in many of his medical colleagues.[69] The Scottish-born doctor is considered one of the founding fathers of evidence-based medicine. A born humanist and empiricist, those traits were sorely tested by all that the tempestuous twentieth century threw at him. Freudian psychoanalysis, for starters. As a medical student, he turned to the talking cure to treat his inability to ejaculate and spent several years in Germany, then Austria, then Holland – following his Jewish therapist as the therapist fled the Nazis.

Cochrane took another year off his studies to fight fascism as a medic in the Spanish Civil War, dodging exploding shells and rival factions and rubbing shoulders in Madrid bars with the likes of George Orwell, Ernest Hemingway and Cochrane's Oxford chemistry professor, J.B.S. Haldane, famed for his 1929 hypothesis on the origin of life on Earth. When Cochrane finally finished his medical degree at University College London, the Second World War broke out. He enlisted in the Royal Army Medical Corps, serving in Egypt and Crete, only to be captured during a failed operation.

Thanks to the fluent German he'd acquired during psychoanalysis, he became the sole medical officer for his fellow POWs in Salonika, Greece, and later in the German camps of Hildburghausen, Elsterhorst and Wittenberg an der Elbe. It turned out to be the only clinical experience Cochrane would ever have. In the German camps his patients had tuberculosis, so he employed the treatment he'd learned at medical school: puncturing one lung, or pneumothorax. Many of his patients died. "I feared I had shortened some lives by doing it on the wrong cases," he wrote in his unputdownable memoir, *One Man's Medicine*.[70]

Liberated from the camps, the broken POW returned to the UK, rehabilitated himself and finally found his spiritual home at the London School of Hygiene and Tropical Medicine. Here humanism and empiricism combined, particularly in the form of Austin Bradford Hill, a medical statistician who carried out the first randomised, double-blinded controlled trial. It was to test the effectiveness of streptomycin in treating tuberculosis. Hill, together with Sir Richard Doll, was also one of the first to prove the link between smoking and lung cancer.

Cochrane employed the lessons he learned from Hill to improve the health of Welsh coalminers afflicted by "black lung" disease (pneumoconiosis), carrying out legendary RCTs that used entire towns to compare treated and untreated populations. Later, as director of the Medical Research Council's epidemiology unit in Cardiff, his team conducted the first RCT to measure whether aspirin could reduce mortality after a heart attack.[71]

Cochrane's reputation saw him invited to give a prestigious lecture in honour of a famous surgeon. It involved writing an accompanying booklet. In Cochrane's own words:

> I ended up writing the book almost entirely between the hours of 10 pm and 1 am ... I date the real beginnings of my love of whisky to this period ... I decided to concentrate mostly on one simple idea – the value of randomised controlled trials in improving the NHS ...

Titled *Effectiveness and Efficiency: Random Reflections on Health Services*, the booklet pulled no punches.[72] Cochrane argued that the National Health Service was not delivering bang for-buck, citing "the minuscule gains in life expectancy" being achieved.[73] He sheeted much of the blame to flimsy treatments that had been rolled out without the backing of evidence. A review in *The British Medical Journal* opined: "The hero of the book is the randomized control trial, and the villains are the clinicians ... who either fail to carry out such trials or succeed in ignoring the results if they do not fit in with their own preconceived ideas."[74]

Cochrane's hastily written text became a manifesto that helped to ignite a global revolution. In Canada, the charge was led by the likes of David Sackett and Gordon Guyatt (who coined the term "evidence-based medicine"). In North America, it was spearheaded by Thomas Chalmers, David Eddy and Alvan R. Feinstein. In Oxford, Iain Chalmers took up the torch. He had also known the anguish of wondering whether he was doing more harm than good during his work in a refugee camp in Gaza from 1969 to 1970, specifically whether he had waited too long to prescribe antibiotics to malnourished children. "It gradually dawned on me there that some things I had been taught in medical school were probably lethally wrong."[75]

Returning to Cardiff to train as an obstetrician, Chalmers was again flummoxed by the conflicting expert opinions around interventions in pregnancy and childbirth. Until, in 1972, Archie Cochrane's "very readable little book" came to his rescue.

Chalmers also found his home at the London School of Tropical

Medicine and Hygiene, and later, at Oxford, helped to organise a systematic review of the evidence around treatments for pregnancy and newborns. It was published in 1989 as *Effective Care in Pregnancy and Childbirth*. One of the treasures buried in the mass of literature was solid evidence for a lifesaving treatment for premature babies. If the mother was injected with a steroid prior to giving birth, it hastened the maturation of the infant's lungs and increased their chance of survival. These results changed medical practice.[76]

Witnessing the persuasive power of the systematic review, in 1992 Chalmers helped to establish the Cochrane Centre at Oxford to apply systematic reviews to all areas of medicine. A year later, the Cochrane Centre combined forces with researchers around the world and became known as the Cochrane Collaboration.

The systematic review, whether by Cochrane collaborators or other groups of researchers, now represents the apex of the pyramid of evidence-based medicine. An understanding of the scientific principles underlying it takes us closer to understanding what went wrong with ivermectin.

The evidence pyramid

How do you start making your way through a mishmash of evidence? You arrange it in a pyramid. The weakest types of evidence get placed at the bottom and the highest at the top.

At the bottom we find studies with no human test subjects involved, such as those in Petri dishes or mice. In the next level up, humans enter the picture through anecdotal evidence such as doctors' case reports. There is "no design" in such studies. Evidence from these bottom two levels can only generate hypotheses, which are then tested in the upper levels of the pyramid. Here we find designed studies. They come in two sorts: "uncontrolled" and "controlled". Uncontrolled studies don't

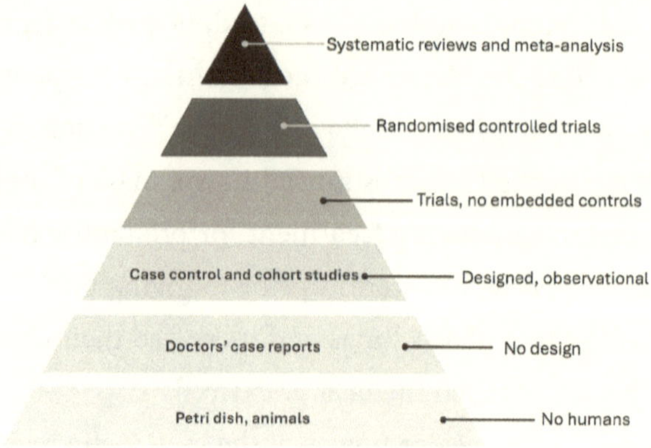

Figure 9. The pyramid of evidence-based medicine.

intervene in people's behaviours. They just observe at people do. Do they or don't they smoke? Do they or don't they take hormone replacement therapy? Hence they're known as observational studies. They may observe retrospectively, as in a "case-control study", or track groups into the future, as in a "cohort study". The effect of an activity is measured by comparing the outcomes of the group who partook in it with those of a carefully designed matched group who did not. Doctors and nurses are favourite choices as study subjects. Observational studies can provide strong evidence if the effect is large, as was the case for the British Doctors Study, which showed that heavy smokers were twenty-five times more likely to get lung cancer than non-smokers.[77] They are harder to interpret if the effect is small, as in the Nurses' Health Study, which looked at the effects of hormone replacement theory on cancer and heart disease in postmenopausal women. The initial finding that hormone replacement therapy was protective against heart disease was not borne out, and the exact risk for cancer is still being determined.[78]

The highest quality of evidence comes from a "controlled trial". Here you treat people more like lab rats, allocating them to two groups, one of which gets the treatment while the other is an untreated "control group".

Controlled trials: a short history

Like many good ideas, it's not new.[79] The ninth-century Persian physician Al-Razi apparently carried out a controlled trial of bloodletting to treat the symptoms of meningitis: "I once saved one group [of patients] by it, while I intentionally neglected [to bleed] another group. By doing that, I wished to reach a conclusion." In thirteenth-century England, the Franciscan monk Roger Bacon – often regarded as England's first scientist – asserted that "without experiment, nothing can be sufficiently known". Even the poets proposed controlled trials. In 1364, in a letter to Boccaccio, Petrarch proposed an experiment to assess whether people would be better off avoiding medical treatment altogether:

> I solemnly affirm and believe, if a hundred or a thousand men of the same age, same temperament and habits, together with the same surroundings, were attacked at the same time by the same disease, that if one half followed the prescriptions of the doctors of the variety of those practising at the present day, and that the other half took no medicine but relied on Nature's instincts, I have no doubt as to which half would escape.

Three centuries later, in 1662, Flemish physician Jan Baptist van Helmont came up with much the same idea:

> Let us take out of the Hospitals, out of the Camps, or from elsewhere, 200, or 500 poor People, that have fevers or Pleuritis. Let us divide them in Halfes, let us cast lots, that one halfe of them may fall to my share, and the others to yours; I will cure them without bloodletting and sensible evacuation; but you do, as ye know … we shall see how many Funerals both of us shall have …

But they were all armchair triallists. The first person we know to have acted on the idea is James Lind. A Scottish ship's surgeon, he published the results of the first controlled trial in his 1753 *Treatise of the Scurvy*.[80] Lind declared his work to be founded "upon attested facts and observations, without suffering the illusions of theory to influence and pervert the judgement".

The trial took place in 1747, while the good ship Salisbury was enforcing a blockade of the English Channel.

> I took twelve patients in the scurvy ... Their cases were as similar as I could have them. They all in general had putrid gums, the spots and lassitude, with weakness of their knees. They lay together in one place, being a proper apartment for the sick in the fore-hold; and had one diet common to all.

Lind divided the group into six pairs and allocated each pair a different treatment for fourteen days. The daily treatments were:

- a quart of cider a day
- twenty-five drops of elixir of vitriol, three times a day
- half a pint of seawater a day
- a nutmeg-sized paste of garlic, mustard seed, horseradish, balsam of Peru, and gum myrrh three times a day
- two spoonfuls of vinegar, three times a day
- two oranges and one lemon a day.[81]

The pair who received oranges and lemons were treated for only six days, because the supply of fruit ran out.

Lind reported:

> The most sudden and visible good effects were perceived from the use of oranges and lemons; one of those who had taken them being at the end of six days fit for duty … The other was the best recovered of any in his condition; and being now deemed pretty well, was appointed nurse to the rest of the sick.

Figure 10. Bloodletting was performed from antiquity into the early twentieth century. In this eighteenth-century Italian engraving, a surgeon is assisted by two attendants.

Unfortunately, the citrus cure did not persuade the navy's "Sick and Hurt Board" for another four decades.[82] Nevertheless, Lind's treatise was the first to articulate the essential features of a controlled trial. The patients were as similar as he could get them and, other than the different treatments, were subjected to the same basic diet and environment. Six decades later, another Scottish physician, Alexander Lesassier Hamilton, conducted a much larger controlled trial involving bloodletting. During the Napoleonic Wars, Hamilton randomly allocated 366 soldiers with camp fever to different treatments at a field hospital: 122 were treated with bloodletting and 244 were not. Of the group subjected to bloodletting, thirty-five died. In the other, much larger group, there were just two deaths. Despite this evidence, bloodletting persisted for another 100 years. The eminent Dr William Osler, who established Johns Hopkins Hospital in Baltimore – a landmark institution of modern American medicine – was still recommending the practice in a 1923 textbook.[83]

So, the controlled trial is not new. What is new is a heightened vigilance against the insidious errors that can skew results – what I have dubbed error gremlins. In the twenty-first century, scientists conducting clinical trials arm themselves against these gremlins with the weapons spelt out in the clunky title of the "gold standard" trial: the large, double blind, randomised, placebo-controlled clinical trial.

Large double-blind randomised controlled trial

As the name suggests, one group of participants (the "control group") will be given a placebo, while the other group is given the treatment being studied. Let's break the other elements down one by one:

Large: The smaller the study, the more the gremlins of chance can muddy the results. Studies should therefore be large, and the overall makeup of the two groups should be generally equivalent as far as age, health, gender, ethnicity, socioeconomic status and so on.

Double blind: Often, both the researchers conducting a trial and those receiving treatment expect a positive outcome. To avoid bias from a placebo effect, both the treaters and the treated are kept in the dark. Neither knows who got the drug or the placebo. Hence the trial is "double blind".

Randomised: Even if participants in a trial are all recruited at the same time and place, how people are allocated to groups could skew the results. For instance, if more sick folk end up in the control group, while more healthy people receive the treatment, it may appear that the treatment worked when in fact the control group was sicker to begin with. To guard against this, some random process, such as generating random numbers on a computer, is put in place to allocate subjects to a group.

Together, these combined elements deliver the gold standard: the large, double blind, randomised, placebo-controlled clinical trial.

In the desperate, early days of the pandemic, some researchers harked back to quainter practices. Hydroxychloroquine, endorsed by President Donald Trump, was the first miracle cure touted for Covid. That was largely because of a trial carried out by flamboyant physician and microbiologist Didier Raoult at Aix-Marseille University.[84] It provides a dramatic example of how the error gremlins can be unleashed by a trial that is small, unblinded and uses non-randomised controls.

None of the safeguards of the large double-blinded RCTs were a feature of the study.

Large? No, the study enrolled forty-two people.

Controlled? Yes, there were untreated controls, but there were problems.

Randomised? No. The controls were from a different hospital and were on average more than ten years older than the treated group (fifty-one years vs thirty-seven years). Since age is a big factor in Covid severity, the untreated controls might have fared worse because of their age.

Blinded? No. The doctors and patients were fully aware of the treatment.

Was the treatment effective? Raoult claimed a 100 percent success rate for hydroxychloroquine because after six days, the virus levels in nasal swabs of treated patients were lower than those found in control patients. But Raoul's round figure excluded three patients who had been treated. One died, one ended up in an intensive care unit and one stopped treatment due to nausea. Moreover, some but not all of those treated with hydroxychloroquine were also given the antibiotic azithromycin, complicating the analysis.[85]

Raoult trumpeted the success of his study until a large RCT showed hydroxychloroquine was more likely to kill than to cure. In January 2025, Raoult's study was retracted by the journal that published it, citing "concerns raised by three of the authors themselves regarding the article's methodology and conclusions".[86]

The large, double-blind, randomised, placebo-controlled trial may be the gold standard. And yet no single trial, no matter how glittering, is enough!

Systematic review and meta-analysis

The idea is simple enough. Dig up all the relevant studies, usually randomised controlled trials (RCTs). Extract their data, pool it and carry out a statistical analysis, the "meta-analysis", to get the best idea of the size and reliability of the treatment in question's effect.[87]

When I first encountered the meta-analysis, it sounded like a terrible idea. How does pooling studies of variable quality take you closer to the truth? It would be like trying to determine the average weight of five-year-olds by including studies that used faulty scales. Rubbish in, rubbish out!

But that's not the idea at all. The philosophy of evidence-based medicine is that even with the best trials, error gremlins may have snuck in. The role of the systematic review and meta-analysis is to flush them all out, wherever they might be hiding. Reviewers will dig systematically and doggedly through mountains of both published and unpublished studies. Finding unpublished studies is difficult but important, because these tend to show unexpected or undesirable effects, especially if they were carried out by drug companies. Leaving them out risks so-called "publication bias", whereby only studies that show the "right" (or desirable) results tend to be published.

While the systematic review aims to unearth every relevant study, it does not lump the good with the bad. Instead, reviewers sort the studies according to their "risk of bias". Studies that meet the "gold standard" criteria outlined above – large studies, with clear randomisation and double-blinding – are considered to have a lower risk of bias.

Next, the body of data is carved up in different ways. How do things look if you only include the high-quality studies? What if you include everything? It's a transparent approach that helps to clarify the true picture of the existing research. Statisticians are then called in to do a meta-analysis. This delivers an overall risk ratio (RR) of the treated participants versus the untreated. An RR of one means the treatment has no effect. An RR of less than one means the treatment lowers the patient's risk. An RR greater than one means the treatment increases the risk. For instance, a meta-analysis for hydroxychloroquine showed that the RR was 1.11, meaning patients who took the drug were at much greater risk than those who received a placebo.[88]

So, now we're ready to take a tour of this evidence pyramid as it applied to the studies of ivermectin and Covid-19.

The ivermectin evidence pyramid

The "irrefutable" studies Ohio lung specialist Fred Wagshul was referring to all lie at the bottom of the pyramid, and in some cases, in the trash can; there were many fraudulent studies.[89] Several wannabe Covid drugs got their start in a Petri dish, including hydroxychloroquine, remdesivir (a one-time Ebola drug), and of course, this is how the ivermectin story began, with Caly and Wagstaff's work at Melbourne's Doherty Institute. Ivermectin, used at over twenty times the recommended dose, eliminated the virus from African green monkey kidney cells growing in a Petri dish. Such a dose was deemed unattainable in people.[90] That is why organisations such as the Gates Foundation and WHO prioritised more promising drugs and did not fund further testing of ivermectin.[91]

The next tier: studies with no design

The main problem with this class of studies is they can't separate correlation from causation. Here's a famous example. It has been observed that when ice cream consumption goes up, so too does the number of accidental drownings.

Perhaps a belly full of ice cream cramps a swimmer's muscles, causing them to drown? It's not unreasonable; that's what I was told as a child. But banning ice-cream consumption would make no difference to the number of people who drowned. The link is not causal. It's just that both ice cream consumption and swimming happen more frequently in summer. In this example, it's easy to spot summer as the hidden variable that drives both increased ice-cream consumption and accidental drownings. Finding the hidden variable is not always so easy.

In early May 2020, the Peruvian government rolled out Mega Operación Tayta (MOT), deploying its army to deliver ivermectin across twenty-five states. In October 2020, a Colombian-based researcher by the name of Juan Chamie, from Universidad EAFIT, reported a strong correlation between the rollout of ivermectin and declining deaths in eight Peruvian states.

The report, titled "Real-World Evidence: The Case of Peru. Causality between Ivermectin and Covid-19 Infection Fatality Rate", appeared on a site called ResearchGate. It looks like an online journal, but the papers here are not peer reviewed.[92]

More than a year later, another version of Chamie's study appeared.[93]

Although it looked more polished and included two co-authors, David Scheim, an "independent medical researcher" based in Virginia, and Jennifer Hibberd, an "instructor" from the University of Toronto, it was still not peer reviewed. For this trio, the correlation between the rollout of ivermectin and declining deaths was enough to claim: "Ivermectin is a compelling option for immediate, large scale national deployments."

But the correlation studies do not actually support that view. For one thing, the authors cherry-picked their locations. "One of the places with the highest death rate, Chincha (in the state of Ica), had the local government distributing ivermectin and still they were hit very hard," Patricia Garcia told me. They also cherry-picked particular Covid outbreaks. For instance, the first Covid wave in Peru receded around September 2020 and correlated with the rollout of ivermectin. The second wave rose in February 2021 and petered out the following May, when there was no ivermectin rollout.[94]

The reality is that waves of Covid infection swept across every part of the world, starting in China, then Europe, South and North America, Asia, Africa and Australia. They continue to do so. Similar waves were seen in the Ebola epidemic in West Africa. They are typical of pandemics.[95] They have nothing to do with ivermectin.

Pierre Kory and his colleagues at the FLCCC also held up these types of correlations as sound "epidemiological evidence". They published a review in the *American Journal of Therapeutics* in April 2021.[96] But they seem to have gotten a few things wrong. A letter to the same journal from Brazilian academics politely pointed out that Kory's Brazilian data "differ from the official data".[97] Moreover, in Paraguay, the number of Covid-19 cases in one state decreased three weeks before the introduction of ivermectin. A year later, the *American Journal of Therapeutics* issued an "expression of concern" over Kory's paper.[98]

There is no doubt that correlations between the rollout of ivermectin and declining deaths are eye-catching. But even if these papers were not riddled with errors, they are scientifically no more conclusive than the correlation between consuming ice cream and drowning. As the Brazilian academics who critiqued Kory's paper put it: "We believe that only large-scale randomized controlled trials can determine the utility, if any, of the administration of ivermectin."

No design-case reports

Rolling out large-scale RCTs is a major production requiring money, manpower and logistical know-how. In the interim, doctors often test the waters by doing an "uncontrolled" study: they treat a small group of patients and report the results, comparing them to similar cases from medical records. That was the approach Sydney-based gastroenterologist Thomas Borody took to test ivermectin. Uncontrolled studies had served him well in the past. In the 1980s, Australian researchers Barry Marshall and Robin Warren made the Nobel Prize–winning discovery that a species of acid-resistant bacteria, *Helicobacter pylori*, caused stomach ulcers. But would eradicating those bacteria eradicate ulcers? To find out, Borody and colleagues recruited eighty-two patients with a history of recurring ulcers and treated them with a cocktail of antibiotics. After four years, they found that those who remained free of the bacteria were cured of their ulcers.[99] At the height of Sydney's Covid epidemic, Borody decided to try another cocktail: ivermectin, zinc and the antibiotic azithromycin.When I interviewed him by phone in August 2021, he told me he'd treated over twenty infected patients of all ages who had contacted him. The youngest was sixteen. Borody was ebullient about the results, telling me that only one patient had died. As far as I can tell, the results of this study have not been published. But Borody has published a similar study conducted in Los Angeles and funded by a Californian company, ProgenaBiome.[100]

Twenty-four people were recruited via doctor referrals or word-of-mouth. Their average age was sixty-six and some appeared to be in bad shape, with oxygen levels well below the norm of 90 percent saturation. Nevertheless, rather than go to hospital, they elected to be treated with ivermectin. They were given the following ten-day treatment:

- ivermectin (12 mg on day one, day four and day eight)
- doxycycline (an antibiotic, 100 mg twice a day)
- zinc (25 mg twice a day),
- vitamin D3 (1500 IU twice a day)
- vitamin C (1500 mg twice a day).

According to the paper, some of those who had been ill for nine days improved within twelve hours of receiving their first treatment. The trial did not include a control group; that is, the researchers did not randomly allocate some of their patients to go without ivermectin. Instead, they compared their results to "background-matched" statistics from the US CDC. The death rate from this data base was around 14 percent; so with twenty-four people, assuming the same death rate, they should have seen three deaths. They saw none. Hence they claimed "an effective and inexpensive Covid therapy preventing hospitalization and death in all patients and rapid resolution of hypoxia in these ambulatory 'moderately to severely' ill patients."

What's wrong with that claim? It's highly susceptible to errors, as are all uncontrolled studies.[101] For starters, it's such a small study that these findings could be due to chance. Toss a coin twenty times and you may well get far more heads than tails. Toss it a thousand times and you see tails and heads are equally likely. Likewise, if the combination treatment had zero effect, enrolling a small number of people means that by chance, it might appear beneficial. Another potential error is bias, conscious or unconscious. To minimise bias and the placebo effect, whereby patients improve because they have a powerful belief in the treatment, the researchers conducting a trial should have a neutral attitude to the outcome, a so-called "clinical equipoise". But in this trial, the authors were hardly neutral. They were so convinced of the benefits of the treatment, they claimed it would be *unethical* to include a control group. Some of the authors of the Borody study also stood to gain financially from the marketing of the "combination therapy". Borody himself had taken out a patent on the treatment.[102]

So the trial operators were not neutral about the outcome of the treatment. What about the patients? Even though they were experiencing very low oxygen levels, they refused to go to hospital. Kurinchi Gurusamy, a professor of evidence-based medicine and surgery at King's College London, wonders whether they were strictly comparable to hospitalised patients in the CDC database. "Clearly these patients felt well enough *not* to go to hospital."[103] The patients were also fully aware they were being given a treatment that the trial operators were enthusiastic about. That means they were perfectly poised for a placebo effect. By some measures, the placebo effect can account for 30 percent of the success of a treatment, although its exact impact is debated.[104] Of course, it's great if patients believe in their treatment, and no one begrudges this extra boost to its success. But what you want is the placebo effect as the *icing on top* of a treatment for which there is solid evidence of benefit. Setting aside a control group among the recruited subjects to receive a treatment lacking ivermectin would have allowed researchers to measure the benefit from the treatment. A third type of error has to do with the accuracy of measurement. Some of these patients had alarmingly low oxygen levels, but those levels were self-recorded using pulse oximeters. The FDA has raised concerns over the accuracy of these meters. As Gurasamy put it, "all these errors were rampant in this study". In August 2021, when I spoke to Borody, I raised the issue of the reliability of his trial and the need for RCTs. His response: "You don't need an RCT to test a parachute." The journal that published Borody's study begged to differ. In 2025, the article was retracted because "concerns were raised regarding methodology used in the study, the patient enrolment process and data analysis."[105]

It's true that an uncontrolled study can give you a clear result when the size of the effect is large. For instance, you don't do an RCT to test the effectiveness of a defibrillator – it's clear the intervention has a massive benefit because the patient lives rather than dies.

Designed observational studies

We'll skip this tier. These types of studies were not significant players in studying the effectiveness of ivermectin against Covid-19.

The penultimate tier: RCTs of ivermectin

Some of the first RCTs carried out for ivermectin showed a strong benefit. On closer examination, however, they fell far short of the gold standard. One study, whose senior author was Nematollah Gheibi from the Shiraz University of Medical Sciences in Iran, found that ivermectin treatment reduced deaths by 15 percent.[106] But it appeared that the trial participants had not been assigned to groups at random. The tell was that the amount of virus measured in the patients differed significantly between the two groups. Had the study operators assigned sicker patients to the control group? If so, then the treated patients would have recovered faster because they were healthier in the first place.[107]

Another RCT made the dramatic claim that ivermectin reduced deaths by 90 percent! The study, reported as a preprint in November 2020, had a huge impact because it was also the largest RCT at that time. The author, Ahmed Elgazzar, a Professor Emeritus at Benha, in

Egypt, had recruited 600 subjects. But a close inspection of the data revealed some problems. For one thing, many of the patients enrolled in the trial had died before the trial began. It was a case of "audacious fraud", noted Kyle Sheldrick, a medical researcher at the University of New South Wales who helped to bring this fraudulent report to the attention of the world.[108] Colombian paediatrician Eduardo López-Medina captured the feeling of many doctors: "I was shocked, as everyone in the scientific community probably were. It was one of the first papers that led everyone to get into the idea ivermectin worked in a clinical-trial setting."[109]

To be sure, some of the first RCTs that reported a beneficial effect for ivermectin were not fraudulent. But those beneficial findings were not borne out. The TOGETHER trial, headed by McMaster University epidemiologist Ed Mills, pulled out every stop to give ivermectin a chance to prove itself. It was carried out in Brazil to catch the peak wave of Covid infections. The low-hanging goal was to see if ivermectin could keep infected people out of hospital, dialling down their illness to the mild form experienced by 90 percent of people. To that end, the researchers recruited participants from the 10 percent of the population most likely to end up in hospital: those aged over fifty; the obese; those with diabetes, hypertension, heart disease, cancer or kidney disease; and transplant recipients.

The other challenge was scouring community clinics to rope in those at risk within the first week of their symptoms, before they got severely ill. And the researchers needed to recruit enough participants for their study to be statistically significant.

They managed to recruit 679 subjects. They were randomly allocated to be given ivermectin or a dummy pill. Neither they nor the trial operators knew which was which. Those given ivermectin received six times the maximum approved dose over three days. The trial operators then tracked whether or not patients were hospitalised within twenty-eight days of receiving treatment. In March 2022 they reported their findings in The *New England Journal of Medicine*. The result? There was no *significant* difference between the treated and placebo groups.[110]

Systematic review and meta-analysis

Systematic reviews and meta-analysis for ivermectin as a treatment for Covid began appearing in 2021. One of the first, published in June, was co-authored by statistician Andrew Bryant at Newcastle University in the UK and Tess Lawrie, the founder of the BIRD Group. Their review of fifteen trials delivered "moderate certainty" evidence that ivermectin reduced the risk of death by 62 percent. If they included "low-certainty" evidence, the risk of death was reduced by a whopping 86 percent.[111]

Two more systematic reviews appeared the following month. One, prepared by statistician Andrew Hill's group at the University of Liverpool, showed a 56 percent reduction in the risk of death.[112] The other, published by the Cochrane Library, found zero effect. Stephanie Weibel and her co-authors concluded: "We are uncertain whether ivermectin ... reduces or increases mortality."[113]

How could three systematic reviews arrive at such different verdicts? They appeared to be playing by the same set of rules, all sifting their studies for "risk of bias" and grading them accordingly. However, one thing the systematic reviewers were not on the lookout for was brazen fraud. That may sound naïve – but as Paul Garner, coordinating editor of the Cochrane Infectious Diseases Group, put it, "everything is based on trust in science". The fishy data was eventually sniffed out by "data detectives" like Kyle Sheldrick at the University of New South Wales and his colleagues.[114]

There were several publications among the RCTs with fraudulent data, some of which have since been retracted.

When Andrew Hill repeated his meta-analysis leaving out the discredited papers, he got quite a different result. "It seems the whole ivermectin story is a house of cards," Hill told me. By way of a *mea culpa,* he penned a lengthy article in January 2022, titled "Ivermectin for COVID-19: Addressing Potential Bias and Medical Fraud".[115]

Punchline for ivermectin

Ivermectin may end up competing with laetrile to be "the most thoroughly studied failure in the history of medicine".[116] The TOGETHER trial (published in March 2022), carried out in Brazil, did find that it had a small, not statistically significant effect on hospitalisation rates, which were slightly reduced. Could this be a tiny signal in the noise? The lead researcher, Mills, suspects that in some Brazilian patients, ivermectin rid them of an underlying parasitic infection and so improved the person's ability to fight Covid. If that's the case, this tiny benefit would be restricted to those parts of the world where parasitic infections are chronic.

To find out whether Western populations, who rarely suffer parasitic infections, might benefit even a tiny amount from ivermectin, the world waited for three large RCTs. The first to appear, in August 2022, was the US-based COVID-OUT trial. Published in the *The New England Journal of Medicine*, it tested the effects of ivermectin as well as metformin and fluvoxamine on 1323 people in the early stage of Covid. The investigators found none of the drugs had any impact on hospitalisation or deaths.[117] A second large US-based trial run by the National Institutes of Health known as Activ-6 (which stands for accelerating Covid-19 therapeutic interventions and vaccines) reported its findings in February 2023 in *The Journal of the American Medical Association.* Tracking 602 people who took a high dose of ivermectin for six days, it showed no protection against hospitalisation or death.[118]

Finally, the PRINCIPLE trial, a large RCT of the British population led by Chris Butler at Oxford, was published in April 2024 in the *Journal of Infection*.[119] It studied 8811 people with Covid, of which 2157 received ivermectin. The trial had previously found that inhaled budesonide, a repurposed asthma drug, was useful on reducing the severity of Covid.[120] Its finding for ivermectin: "Ivermectin for COVID-19 is unlikely to provide clinically meaningful improvement in recovery, hospital admissions, or longer-term outcomes."

What can we learn?

Imagine being an emergency room doctor in the early days of Covid. You're seeing patients who days before were fit and healthy or at least looking forward to many more years of life. Suddenly, they're dying in a way that nothing in your training has equipped you for: they are asphyxiating from pneumonia, or their bodies are shutting down because their organs have failed. If there's an off-the-shelf drug that offers a ray of hope, why not try it?

And you do. But it needs to be done as part of a clinical trial. Otherwise, you don't know if you've done more harm than good.

Ivermectin used as a Covid treatment caused harms. Some were clearly visible, such as in the Native Peruvians with necrotic ulcers on their arms, and the people reporting to emergency rooms all over the world with ivermectin poisoning. But there were less measurable harms: all the people who dosed up with ivermectin and went about their daily business in the false belief they were protected. Some combination of these factors may explain Peru's extreme death rate early on in the pandemic.

As ivermectin fever spread to the West it created new havoc, offering itself as an alternative to vaccines. There are stories of true believers who rejected vaccination and died from Covid while still clinging to their ivermectin cure.[121]

Getting the ivermectin RCTs underway sooner might have helped. Hydroxychloroquine mania was tamped down within months by fast-moving, rigorous RCTs. The large, robust RCTs for ivermectin didn't report formally until 2022 for Brazilian populations, and until 2023 and 2024 for US and British populations. There were reasons for the delays: the low likelihood of success given the unattainably high doses required, and the difficulties of recruiting enough patients for the studies. Nevertheless, the delays caused harm. Misinformation thrives in a vacuum. Conspiracy theorists co-opted ivermectin into a narrative that the world's institutions, big pharma and scientists were supressing ivermectin in favour of commercially lucrative vaccines and new drugs.

But now we have the data. So perhaps the ivermectin debacle is just an unfortunate aberration that we can put down to the panic and chaos of a pandemic?

That doesn't seem to be the case. Ivermectin has become entrenched as part of a narrative of medical freedom championed by some doctors' groups.[122] Notwithstanding there is not a single clinical trial to support its use, the US Front Line COVID-19 Critical Care Alliance (now the Independent Medical Alliance) now promote ivermectin for treating flu and respiratory syncytial virus. US Republicans have also embraced the ivermectin cause. In May 2022, they pushed legislation to promote ivermectin as a treatment for Covid and protect doctors who prescribe the drug to patients.[123]

Robert F. Kennedy Jr is one of those who claim ivermectin and hydroxychloroquine were discredited because they threatened a billion-dollar "vaccine enterprise".[124] And that fewer people would have died if the United States had deployed these drugs for Covid. "There's ... countries that did the opposite of what we did – that provided ivermectin, hydroxychloroquine, other early treatments to their populations – and had 1/200th of our death rate", he told Fox News in July 2023. He also seemed to claim that the virus was designed to spare Jews and Chinese. "Covid-19 is targeted to attack Caucasians and Black people. The people who are most immune are Ashkenazi Jews and Chinese.[125]

"For two decades, Kennedy has been arguably the world's most prominent anti-vaccine activist.[126] He chaired the Children's Health Fund, which spreads disinformation about vaccines, and visited Samoa to support the anti-vaccine movement months before a 2019 measles outbreak that resulted in eighty-three deaths, mostly children. "The Samoan incident showed us how disinformation can kill," said Paul Offit, a paediatrician and infectious disease expert based at the University of Pennsylvania. "He sowed further distrust, he jumped all over it – he met with anti-vaxxers in Samoa to promote the notion that 'it's not measles, it's the vaccine', and immunization rates dropped."[127]

As I pen the final words to this chapter, Kennedy is now the Trump-appointed secretary of the US Department of Health and Human Services. He oversees everything from the Centers for Medicare and Medicaid Services to the vaunted institutions that protect Americans against disease outbreaks and evaluate new treatments: the Centers for Disease Control and Prevention (CDC), the National Institutes of Health (NIH) and the Food and Drug Administration (FDA).

Measles, the most contagious disease known, is surging in the US as a result of vaccine hesitancy. As of 8 May 2025, a total of 1001 confirmed measles cases, most in unvaccinated children, were reported by thirty states. One hundred and twenty-six were hospitalised, and three died – the first measles-related deaths in the United States in a decade.[128]

At the FDA, Peter Marks, who oversaw the safety and efficacy of vaccines and was instrumental in the implementation of Trump's warp-speed program to develop a Covid vaccine, "resigned" on 29 March. He was told he'd be fired if he didn't.[129] His resignation letter described his attempts to address the concerns of the Health and Human Services secretary regarding the safety of common childhood vaccines. "However, it has become clear that truth and transparency are not desired by the Secretary, but rather he wishes subservient confirmation of his misinformation and lies." He continued, "My hope is that during the coming years, the unprecedented assault on scientific truth that has adversely impacted public health in our nation comes to an end ..."

4

Human Consciousness

Human consciousness is just about the last surviving mystery.

Daniel Dennett[1]

If Karl Popper believed that scientists were engaged in "disproving" their theories, Thomas Kuhn believed the opposite. Most scientists, blinkered by a "dominant paradigm", worked to shore up that paradigm rather than challenge it. Kuhn referred to this as "normal science". Only when the dominant paradigm was shown to be fatally flawed – usually death by a thousand cuts – would a scientific revolution occur. Kuhn would have found much to support his philosophy in the convulsions experienced by the new science of consciousness.

Making consciousness respectable

British physicist Francis Crick liked to solve other people's problems.[2] Annoyingly so: his Cambridge colleagues would hide their puzzles from him.[3] But Crick's inquisitiveness paid off. In 1953, aged thirty-seven, while working on his PhD thesis, together with James Watson, an American postdoctoral fellow twelve years his junior, he solved the structure of DNA. In so doing, he helped to unravel the mystery of how living organisms replicate their genetic blueprint. It won Crick and Watson the Nobel Prize. Having cracked the mystery of life, Crick later turned his talents to the other big mystery of the universe: consciousness. (When he was a young man, his mother once asked him what scientific problems he wanted to pursue in life. Francis replied that there were only two problems that interested him: the mystery of life and the mystery of consciousness.[4])

It's not uncommon for scientists to pivot this way in their senior years – it has been dubbed the "philosophase". What distinguished Crick was his success in placing consciousness, for 2000 years the stomping ground of philosophers, within the remit of measurable, testable science. As Leslie Orgel, a colleague at the Salk Institute, put it, "Because of his stature, he made brain science respectable."[5]

Consciousness had not previously been deemed a respectable subject for science. It was "generally regarded as a form of bad taste, and graduate students, who are always attuned to the social mores of their disciplines, would roll their eyes at the ceiling and assume expressions of mild disgust," noted American philosopher John Searle.[6]

Psychologists had disavowed it, precisely because it could not be measured or tested. "[I]t is impossible to specify what it is, what it does, or why it evolved. Nothing worth reading has been written on it," stated the 1989 edition of *The International Dictionary of Psychology*.[7]

Some held to the view expressed by nineteenth-century German neurophysiologist Emil Du Bois-Reymond: "ignoramus et ignorabimus" – we do not know and we will not know. For Crick, that was like a red rag to a bull. His credo was "In any field, find the strangest thing and then explore it", as urged by physicist John Archibald Wheeler (whom we will meet again in the next chapter).[8]

In 1977, at the age of sixty-one, Crick moved to California's Salk Institute and began attracting converts to the cause. The most ardent was Christof Koch, forty years Crick's junior, an exuberant American computational neuroscientist (someone who uses mathematical models to study the brain) based at the California Institute of Technology (Caltech).

Crick and Koch's first paper, published in 1990, was titled "Towards a Neurobiological Theory of Consciousness".[9] It was something of a manifesto. Surprisingly, they did not attempt to define consciousness. Many scientists would say that a clear definition of the question is the starting point of a scientific study. But Crick and Koch demurred.

> We feel that it is better to avoid a precise definition of consciousness because of the dangers of premature definition. Until we understand the problem much better, any attempt at a formal definition is likely to be either misleading or overly restrictive, or both.

The definition of consciousness remains problematic. Philosopher Thomas Metzinger referred to it as an "ill-defined folk psychological umbrella term".[10] And at a recent conference of consciousness researchers, neurologist Joseph LeDoux quipped: "There are about 700 definitions in this room."[11] In the absence of a clear definition, there were risks. For a start, how could researchers be sure they were investigating the same thing?

Nevertheless, the history of science shows that progress can be made in the absence of an agreed definition. The exact nature of heat, for instance, was a mystery until the seventeenth century. Some considered it to be like a fluid that could flow from one body to another. Others (correctly) inferred it to be an inherent property of moving atoms. Progress did not rely on defining heat but on measuring it, which was made possible by the invention of thermometers.[12]

Crick and Koch proposed a similar strategy for consciousness: they would try to measure something, even if there was still disagreement about what consciousness was. More specifically, they would measure the brain activity that corresponded to a conscious experience, or, as they put it: "the neural correlates of consciousness".

The time was ripe for such investigations. In 1990, the same year Crick and Koch published their manifesto, technology delivered a marvel. Previously, the thick human skull had rendered the workings of the brain invisible. But as French neuroscientist Stanislas Dehaene put it, "in 1990 the skull became transparent", thanks to functional MRI (fMRI).[13] Like a movie camera for the mind, it tracked the frenetic activity of the brain, albeit indirectly. Not by following the electrical chatter of brain cells, but by tracking the increased blood flow required for that activity.

As well as this new brain camera, Crick and Koch built on decades of work by experimental psychologists who had developed model systems to try to lay bare the experience of consciousness, isolating it from other mental processes. A few examples will make this crystal clear.

Figure 11. The limited theatre of consciousness. Is it two faces or a vase? At any one moment, only one perception is possible.

One was "blindsight". First observed in the 1970s, a person with blindsight has no conscious ability to see. But if you ask them to walk through a room, they won't bump into the table or chairs. They are receiving visual images of the furniture, without being conscious of it. Blindsight results from damage to part of the visual system at the back of the brain. But you don't need a brain-damaged subject to study this disconnect between the sensory input a person receives and what they are conscious of. Optical illusions are a prime example of this schism. One moment you're looking at a vase; the next it appears to be two silhouettes. You haven't altered your gaze, but something shifts in what you're conscious of seeing. (See Figure 10.)

Or there's the famous Harvard gorilla experiment.[14] Subjects were asked to watch a video showing six basketball players passing balls to one another – three clad in white shirts, three in black. The viewers were asked to count the number of passes between players in white shirts. Midway through the exercise, a man dressed in a gorilla suit sauntered onto the court, stood briefly in the middle facing the camera, beat his chest, and then moved on. When the viewers were asked what they saw, half reported the number of ball passes between white-shirted players and nothing else. They had not seen the gorilla! It's a stunning example of "inattentional blindness" – what you're focused on influences what you're conscious of.

A key tenet of Crick and Koch's manifesto was to confine their exploration of consciousness to vision. They hoped that whatever they learned here would generalise to other facets of consciousness, such as pain, emotion, thinking or the self. Moreover, when it came to understanding brain circuitry, the wiring of vision was among the best understood. Patterns of light falling on the retina excite wiry brain cells known as neurons. The electrical signals funnel into the optic nerve at the back of the eye and from there to the egg-shaped thalamus, a routing station at the centre of the brain. From there, the visual signals are relayed to the back of the brain, to the visual cortex, a region that occupies nearly a fifth of the wrinkled outer layer of the brain.[15] Here, over a dozen brain processors detect specific features such as edges, orientation, colour and movement. The output of this visual neural network builds up the final dynamic picture that we see.

The "hard problem"

Crick and Koch's manifesto launched a new era of discovery for both scientists and philosophers. One of the recruits was David Chalmers, an Australian philosopher of mind, then based at the Australian National University.

Although inspired by the scientific mission, his 2000-year philosophical tradition gave him pause. In particular, the arguments of philosophers such as seventeenth-century German philosopher and mathematician Gottfried Leibniz. Leibniz, who invented calculus and mechanical calculators, tried to imagine the whirring machinations of the mind as resulting from some kind of machine, perhaps a windmill:

> And, supposing that there were a mechanism so constructed as to think, feel and have perception, we might enter it as into a mill. And this granted, we should only find on visiting it, pieces which push one against another, but never anything by which to explain a perception. This must be sought, therefore, in the simple substance, and not in the composite or in the machine.[16]

In other words, if you scrambled inside a windmill and saw the pulleys and gears, all would be revealed about the mechanism that enables grindstones to grind grain. But if you were to enter a brain and see "pieces which push against each other", how would that shed light on the workings of a mind?

For seventeenth-century French philosopher René Descartes, the stuff of minds and the stuff of machines were totally different: the mind was composed of *res cogitans* (thinking stuff) while the material world was composed of *res extensa* (extended stuff). This "dualist" view remains compelling for some modern-day philosophers such as Chalmers, and even for some scientists, as it was for Nobel Prize–winning neuroscientist John Eccles.[17] Whatever cool toys science might deliver, an "explanatory gap" would yawn, unable to build a bridge to the workings of the mind.

In 1995, Chalmers coined a neologism for the new generation of consciousness researchers. They might well track down the brain mechanisms that correlate with consciousness, but that would be *the easy problem*. Explaining how those mechanisms create the ineffable experience of the colour red – the so-called *qualia* of red – would remain "the hard problem".[18] The hard problem looms large in any discussion of consciousness, but it's also hotly debated.

As for the easy problem (finding the neural correlates of consciousness), Chalmers was willing to bet even that would *not* be easy. In 1998, at the second annual meeting of the Association for the Scientific Study of Consciousness (ASSC) held in Bremen, Chalmers met Christof Koch. The philosopher and the computational neuroscientist hit it off and spent the night crawling the pubs of the historic German town. In the wee hours of the morning, they made a wager: the neural correlates of consciousness would be nailed within the next twenty-five years. Koch bet in favour; Chalmers against.[19]

Twenty-five years later, at the 2023 meeting of the ASSC meeting in Greenwich Village, New York, I reported the result of that wager for the US magazine *Science*. Forgive me, but I must keep you in suspense. Before the big reveal, we need to take a look at what transpired in the intervening twenty-five years. How did scientists go with their brain cameras and experimental tricks for nailing the neural correlates of consciousness?

Hard problem, easy problem, real problem, explanatory gap, *qualia*

American philosopher Daniel Dennett, who passed away in April 2024, battled David Chalmers over the "hard problem" for decades. For Dennett, the idea is a philosophical throwback that answers a scientific problem by invoking mystical forces. Life, for instance, was held to be due to a mysterious *élan vital* until DNA and the genetic code revealed its blueprint and machinery. As Dennett put it, Chalmers' hard problem "is just vitalism reborn".[20] For Dennett, consciousness does not require "an extra soupçon of something". It will emerge once the cogs and pulleys of underlying brain processes are revealed. British neuroscientist Anil Seth has dubbed this quest to fully understand the mechanisms that generate consciousness "the real problem".[21]

Dennett also has little use for the idea of *qualia* as some ineffable property of consciousness that defies a functional analysis. "I think in the end the idea of qualia is broken; it means too many different things to different people; it's a philosophical term of art [that has] never been well defined." He cites the work of musicologist and neuroscientist David Huron, who analysed the qualia of musical tones. "It's not ineffable if you do the science."[22]

A deluge of data and an embarrassment of theories

Crick and Koch's manifesto lured researchers in, and a deluge of data ensued. But science does not progress by data alone. After all, the goal of science is to explain the world – and to do that, you need hypotheses and theories.

What's the difference? A hypothesis is a starting point, a best guess. You carry out experiments to test your hypothesis. If the data support it, the hypothesis solidifies into a theory. For instance, if my car engine fails to kick over, I'll hypothesise that my battery is flat. If I charge the battery and my car starts, then my working theory is that whenever my car fails to start, it's because of a flat battery.

When Crick and Koch penned their manifesto in 1990, they had a hypothesis for what the neural correlates for consciousness would look like. Paying attention to something would trigger a group of neurons across many parts of the brain to fire their electrical signals in synchrony, like a humming choir.[23] "This attentional mechanism helps sets off the relevant neurons to fire in a coherent semi oscillatory way … so that a temporary global unity is imposed on neurons in many different parts of the brain. These oscillations then activate short-term (working) memory."[24] Note this hypothesis suggests that attention and short-term memory are not integral parts of the neural correlate of consciousness, but closely associated with it. Try to remember this hypothesis as we work our way through what transpires in the ensuing three decades. It remains very much alive, though now as part of a theory that Koch rejects.

By 1996, Crick was bullish that the seat of consciousness would be nailed by the end of the century.[25] A big part of the reason was a powerful new model to tease vision apart from conscious perception. Neuroscientist Nikos Logothetis of the Max Planck Institute for Biological Cybernetics had just published a paper describing "binocular rivalry". A human brain normally has no trouble combining the slightly different images it receives from the left and right eye. But if very different images are shown to each eye, say a house to the left eye and a face to the right eye, and if they seem to originate from the same place, there's rivalry: a house is seen, then a face is seen and so on, back and forth, as they wrestle for dominance. The visual input remains the same, but conscious perception is shifting. So what is happening in the brain while this is going on? Logothetis found that the signals in the brain's visual centres remained largely unchanged. The major shifts were occurring elsewhere. Exactly where remained to be determined, but here was clearly a chance to find the lair of consciousness.

For Crick, the new studies of blindsight and now binocular rivalry were "among the opening salvos of a concerted attack on the baffling problem of consciousness". He concluded:

> No longer need one spend time attempting to understand the far-fetched speculations of physicists, nor endure the tedium of philosophers perpetually disagreeing with each other. Consciousness is now largely a scientific problem. It is not impossible that, with a little luck, we may glimpse the outline of the solution before the end of the century.

Crick died in 2004 at the aged of eighty-eight, his zeal undiminished to the last. In his hospital bed he worked on a paper proposing that the claustrum, a tiny structure beneath the cortex, might fit the description for a neural correlate of consciousness.[26]

Fast-forward twenty years. With ever more dazzling cameras to peer into the workings of the brain, there has been an avalanche of data. Some twenty different theories of consciousness, each with its own camp of supporters, purport to explain these data. That's not supposed to happen. In the face of the data avalanche, the flimsier theories should have been flattened, or as Popper put it, "falsified". But no overarching paradigm of consciousness had yet emerged. As New York University philosopher Ned Block told me, the field is "pre-paradigmatic".[27]

Liad Mudrik diagnoses the problem

The scientific machinery appeared to have stalled. Liad Mudrik, based at Tel Aviv University, helped diagnose the cause of the problem. No doubt her training as a philosopher and a neuroscientist helped – she holds PhDs in both . She had also spent two years in Christof Koch's lab at Caltech.

Mudrik wondered: were the scientists acting like Popperians or Kuhnians? To find out, her student Itay Yaron analysed over 400 recently published experiments on consciousness. Most made no attempt to challenge any of the theories. When they did set out to test a particular theory, 80 percent of the time the experiment ended up supporting it. No papers refuted any theory. The behaviour of consciousness researchers thus appeared to be textbook Kuhn. Most telling was that the results depended on what type of experimental test the researcher was using. In particular, if the test required the subject to "report" when they were conscious of a particular object, say a gorilla, the results were more likely to favour one theory.

If the test detected conscious perception without requiring the subject to report it, the results favoured a different theory. One way to bypass the need for a report relies on binocular rivalry. A screen can be set up with a divider so that one of the subject's eyes is shown a pattern of stripes moving to the right and the other a pattern moving to the left. The subject will alternate between tracking one or the other pattern, thus indicating which pattern they are conscious of at that moment.[28]

In 2022, Mudrik and Yaron published a paper on their findings with two other colleagues, experimental psychologists Lucia Melloni from the Max Planck Institute and Michael Pitts from Reed College, Portland. As it happened, Mudrik, Melloni and Pitts were already hard at work, trying to kick-start the stalled scientific machine.

...

The trio had met back in 2018, when they accepted an intriguing invitation to a weekend in Seattle. It was to be hosted by Christof Koch, by then the director of the Allen Institute for Brain Science (established by Microsoft founder Paul Allen) and Dawid Potgieter, the energetic new director of science programs at the Templeton World Charity Foundation (TWCF).

A signature of projects funded by the Templeton Foundation is that they cross traditional boundaries, particularly those between religion, philosophy and the various outposts of science.[29] With an Oxford PhD in neuroscience under his belt, Potgieter viewed the scientific study of consciousness as a perfect fit for the Templeton Foundation. But as he researched the field, he was surprised to find it overgrown with gnarly theories. How could such a thicket be pruned?

Aggressively, suggested Koch, by way of an "adversarial collaboration". The key attribute of a successful theory is that it should make testable predictions. But as we've seen, each camp was designing experiments to affirm rather than challenge the predictions of their theory. The idea of the adversarial collaboration was that by co-designing an experiment, the camps would put their competing predictions to the test. One or the other should emerge the winner.

There were precedents. Koch pointed to a rather glorious one in physics, when Arthur Eddington tested Newton's and Einstein's rival theories of gravity in 1919. That test proclaimed Einstein's theory victorious (see Chapter 5 for more). A somewhat less glorious precedent came from Israeli-American psychologist Daniel Kahneman. Oddly for a psychologist who never took an economics course, he won the 2002 Nobel Prize in Economics.[30] His "prospect theory", which analyses how people make decisions, gave birth to the discipline of behavioural economics. One of Kahneman's other claims to fame is that he coined the term "adversarial collaboration". In the 1980s, he engaged in one to solve a marital feud. He and his wife, psychologist Anne Treisman, found themselves in vehement disagreement about a theory that explained how people perceive motion. Kahneman proposed they engage in an adversarial collaboration to resolve their feud. As he described it, "people who don't agree on a scientific idea commit themselves to work together towards a joint truth, either by experimentation or by discussion".[31] Their adversarial collaboration didn't resolve their differences. Nevertheless, Kahneman remained a champion of adversarial collaborations. In fact, the last one he championed, before his death in 2024, is the one you are about to hear about.

A weekend in Seattle

Koch's idea for an adversarial collaboration struck a chord with Potgieter. And so, in early 2018, twelve consciousness researchers from across the globe – neuroscientists, psychologists and philosophers – found themselves in receipt of an intriguing invitation to a weekend in Seattle. Among the neuroscientists were three proponents of competing theories: Stanislas Dehaene, the architect of the Global Neuronal Workspace Theory (GNWT), Giulio Tononi, architect of Integrated Information Theory (IIT) and Hakwan Lau, a leading champion of Higher-Order Theories (HOTs). Amongst the philosophers was David Chalmers of 'hard problem' fame, based since 2012 at New York

University. Their goal was to explore the possibility of an adversarial collaboration to test theories of consciousness.

Mudrik, Melloni and Pitts had been invited because they were unaligned to any of the theoretical camps. They also had an unusual history of publishing papers that challenged theories of consciousness. On the Saturday afternoon in Seattle, each had found themselves at the whiteboard, drawing up schemes for discriminating between the predictions of the different theories. "At some point, I think Giulio suggested, 'Why don't the three of you lead the project?'" Pitts recalled. "We had no idea what we were in for. It's consumed our lives."[32]

Over the next nine months, the trio led further discussions around the whiteboards at the labs of the adversaries. The theorists drilled down into their theories and offered new predictions – one of the highlights of the collaboration. Mudrik was impressed by the willingness of the adversaries to negotiate. "It takes a lot of courage; you're putting your neck on the line," she told me.[33] Finally the team came up with an experimental design for disentangling the predictions of two of the theories: IIT and GNWT. The third, HOTs, was left for a different adversarial collaboration.[34]

It would take over four years and involve laboratories around the world to put the ambitious new experiment to the test. The hope was that the result would move consciousness research forward, along the lines of Arthur Eddington's dispositive experiment, rather than Daniel Kahneman's inconclusive one.

The result of the first adversarial collaboration, to be described in this chapter, concerns only GNWT and IIT. But there are at least twenty theories of consciousness, some of which are being tested in other Templeton-funded adversarial collaborations.[35] Let's take a brief tour of some of these theories, before zeroing in on the contestants of the first tournament.

Theories of consciousness draw inspiration from diverse sources, ranging from classical philosophy to quantum physics.[36] Arguably, every one of us is an expert if we accept the definition of consciousness proposed by philosopher Thomas Nagel – he defines it as the experience of being you.[37] What is it like to be Elizabeth Finkel? Only I can say.

Most contemporary theories of consciousness are grounded in brain circuitry. Two are not, and they capture the popular imagination. One is quantum consciousness, proposed in the mid-1990s by Nobel Prize–winning British physicist Roger Penrose (the Nobel was for his part in predicting the existence of black holes). Penrose's theory holds that consciousness relates not to the circuitry of neurons butto the quantum states of tiny structures (microtubules) within neurons. Christof Koch, collaborating with other researchers, including a group from Google Quantum AI, has recently proposed a way to put quantum consciousness to the test.[38] Penrose's theory is reminiscent of physicist Erwin Schrödinger's 1940s hypothesis that genetic information and mutation would be explained by quantum states with an "aperiodic crystal" and likely involve "other laws of physics".[39]

The second theory untethered from brain circuitry is panpsychism, which holds that fundamental particles of consciousness permeate the universe. It also requires other laws of physics. Panpsychism offers no test of its predictions, and so it has no place in science. For Dennett, it has no place in philosophy, either. He described it as "almost embarrassing".[40]

To enter the arena of the Templeton-funded adversarial collaborations, a theory has to make testable predictions. Besides GNWT and IIT, three more have made the grade. I'll hazard a brief description of them here.

First Order Theories (FOT), also known as localist theories, hold that consciousness is primarily generated by looping circuits in the sensory regions of the brain. As philosopher Ned Block told me, "Perceptual consciousness is processed in the areas that do the perceptual analysis."[41] Like a high-level presidential briefing, these representations are delivered to the executive centre of the brain to be acted upon. These theories, like IIT, predict that consciousness is located in the rear, sensory region of the brain.

Higher Order Theories (HOT), which go back to philosophers including John Locke and Immanuel Kant, hold that processing in the sensory regions is just the first step; a higher-order "meta" representation is also required. That higher order "entails some kind of minimal inner awareness of one's ongoing mental functioning", or a "knowing about knowing". In plain language, this seems to equate to a stripped-down version of self-awareness. As Hakwan Lau, who championed HOT in a Templeton tournament, explained to me: "It is a very minimal notion of 'self', just meaning that some part of a system monitors another part of a system … it is more like the kind of 'self' that a self-driving car has."[42] Like GNWT, HOT propose that the frontal region of the brain is crucial for consciousness.[43]

Lastly, Predictive Inference (PI) theories view consciousness as the brain's prediction machine. Ambiguous sensory signals are constantly pinging the brain, so the task for consciousness is to weigh the strength of those signals in order to make its best guess. Is the scary shadow a craggy tree or a malevolent stalker? Optical illusions show that sometimes our consciousness can't decide. Schizophrenia, similarly, might be explained by a predictive glitch that makes sufferers err on the side of hallucinations.

That's it. We're ready to profile the two contestants of the first Templeton-funded adversarial collaboration. Now, maximum brain power please …

Global Neuronal Workspace Theory (GNWT)

The first of our two contenders is Global Neuronal Workspace Theory. "Workspace" refers to a feature of artificial intelligence systems developed in the 1970s to decode speech. While programs on a computer don't generally interact – they run in parallel – the workspace was an exception. Here, independent "knowledge sources" interacted to compare different hypotheses.[44]

In the 1990s, psychologist Bernie Baars at the Neurosciences Institute in San Diego saw an analogy between these AI systems and the processes at work in our brain.[45] Unconscious processes were analogous to computer programs running in parallel. They also represent the lion's share of your brain's activity. No need to think about controlling the beating of your heart, or how to put together the discombobulated edges, shapes and colours that create visual images. And once you've learned to do it, there's no need to think about how to walk or cycle or drive or carry out simple arithmetic. Even complex problems benefit from unconscious brain work. A chess master need only glance at a board to divine their next move. While few of us are chess masters, who hasn't had the experience of stewing on a problem, only to have the answer served up a day or so later from the depths of your unconscious mind? My favourite example is the nineteenth-century German chemist August Kekulé. Chemists struggled for decades to deduce the structure of benzene, an oily component of petroleum. They knew it was made of six carbon and six hydrogen atoms. But how to solve the structure of this twelve-atom puzzle? In 1865, Kekulé dreamt of a snake devouring its tail – the ancient Egyptian symbol known as the ouroboros. It gave him the answer. Benzene was a ring!

According to Baars, if these feats of the unconscious brain were analogous to AI programs running in parallel, then consciousness was something we experience when they interact in a workspace. Another feature of the AI workspace was that while there was no limit to the number of independent programs that could be run, the workspace was limited. So too with consciousness. While our unconscious mind churns away, most of us have trouble focusing on two things at once, as some venerable philosophers have noted. "Ideas impede each other," observed Descartes in 1648.[46] Or as Hippolyte Taine, another French philosopher, put it in 1870, we are furnished with a limited "theatre of consciousness", a metaphor not dissimilar to Baars' workspace.

> You may compare the mind of a man to the stage of a theatre, very narrow at the footlights but constantly broadening as it goes back. At the footlights, there is hardly room for more than one actor ... As one goes further and further away from the footlight, there are other figures less and less distinct as they are more distant from the lights. And beyond these groups, in the wings and altogether in the background, are innumerable obscure shapes that a sudden call may bring forwards and even within direct range of the footlights. Undefined evolutions constantly take place throughout this seething mass of actors of all kinds, to furnish the chorus leaders who in turn, as in a magic lantern picture, pass before our eyes.[47]

Taine's spotlight and Baars' workspace are analogies for the bottleneck that restricts the access to consciousness.

Psychological hacks can reveal that bottleneck. Focus on counting basketball passes, and you're blind to a gorilla in plain view. Or view a face on a screen for less than a twentieth of a second and, if it's followed by odd shapes (a technique known as masking), you'll swear it wasn't there.

In the late 1990s, French neuroscientist Stanislas Dehaene, together with Jean-Pierre Changeux and Lionel Naccache, all based at the

Collège de France, began fleshing out the workspace theory with new data from brain imaging experiments.[48] Two decades on, Dehaene is the leading protagonist of the theory. In his view, the conscious workspace is not associated with a region of the brain, in the way that emotion is associated with the amygdala. Rather, as Crick and Koch hypothesised in 1990, it is a coherent signature flitting across the otherwise noisy brain circuitry.

Our brain is a network of around 90 billion neurons; at any one moment, millions of them are firing off electrical signals. Like tweets crisscrossing the networks of the planet, most trigger no response. But just as a catchy tweet is occasionally retweeted enough to trigger a twitterstorm, so too a particular brain signal will trigger a chorus of "synchronised neuronal oscillations" across the brain. Dehaene believes something like this is taking place during a conscious experience.[49] A lot of sensory input comes in that we don't register, like the invisible gorilla, because the neurons tweeting "gorilla" are getting no response. But if their signal gets massively retweeted, they cross the threshold into the workspace and suddenly there's the big guy, right in front of your eyes. For Dehaene, this twitterstorm of neurons broadcasting their information across the brain *is* the experience of "consciousness".

Over the last two decades, several distinct patterns in brain waves have been hypothesised as features of this twitterstorm.[50] For the purposes of the Templeton tournament, two key features were predicted. One was that the onset of the twitterstorm, described as the "ignition of the workspace", would be detected around a third of a second after the exposure to the stimulus. The other was that the workspace involves the front of the brain.

Integrated Information Theory (IIT)

GNWT's rival at the Templeton tournament, Integrated Information Theory (IIT), was first articulated in 2004. It is primarily the brainchild of Italian-born neuroscientist and psychiatrist Giulio Tononi, based at the University of Wisconsin–Madison.[51] Tononi had spent many years collaborating with Gerald Edelman, a Nobel Prize–winner in immunology who, like Crick, experienced a late "philosophase" and moved on to the problem of consciousness. Like philosopher David Chalmers, Edelman and Tononi were concerned about the "hard problem". They came to the view that decoding brain circuitry would not build a bridge to explaining consciousness. Their solution to the hard problem was to circumvent it, by not starting with any kind of circuitry at all. The theory has captivated several of the diehards of consciousness research, including Christof Koch. "What convinced me, finally, about IIT, is it started with phenomenology", Koch told me during a Zoom call in May 2023.[52]

Before we launch in, here's a warning. Many folks, myself included, have difficulty understanding this theory. This includes neuroscientists, philosophers and a hardcore computer scientist.[53] But as Lucia Melloni told me, "Just because a theory is hard to understand doesn't make it wrong."[54] Indeed, supposedly fewer than twelve men understood Einstein's theory of General Relativity, according to a *New York Times* headline in 1919. According to Arthur Eddington, it was just two – himself and Einstein! (Asked if it was true that just three people in the world understood the theory, he is said to have replied, "Who's the third?"[55]) And when it comes to the physics of the subatomic world, "nobody understands quantum mechanics", confessed physicist Richard Feynman.[56]

So, with that caveat, here's my best shot at describing IIT.

IIT rejects analogies of consciousness to any kind of circuitry – silicon or neural. It starts with the phenomenon of consciousness itself. It describes five axiomatic properties of the conscious state, puts forward postulates about the physical structure that would display such properties, and captures them in a mathematical model.

The five axiomatic properties of consciousness are held to be: intrinsic existence, composition, information, integration and exclusion.

Intrinsic existence refers to the subjective reality of the conscious experience.

Composition refers to the well-structured nature of consciousness. When you look at a face, you see a structured whole – da Vinci's *Mona Lisa* rather than Picasso's jumbled 1910 *Portrait of a Woman.*

Information refers not only to the information carried within a conscious state, but also to the information it leaves out. It is information-rich.

Integration means that every experience is unified rather than reducible to components.

Exclusion refers to the uniqueness of each experience; it is exclusive of other experiences.

From these axioms come postulates about the physical structure that best accounts for consciousness – what sort of "windmill" it must be to produce these five qualities. IIT holds that the physical structure of consciousness is one that is optimised to make a difference to itself, and one which cannot be reduced to its components. To find the signature of such a structure in the brain requires looking for something that demonstrates these characteristics – what proponents of this theory call a "maximally irreducible cause–effect structure".

In Tononi's words: "So, within your brain for instance, the claim would be that there is a set of neurons right now that, as a system, makes the most of a difference to itself in the past and in the future ... It's the most irreducible form that says the past must have been that particular way and the future must be that particular way."[57]

Feeling confused yet? British neuroscientist Anil Seth, based at the University of Sussex, has also struggled to comprehend IIT. He likens the maximally irreducible cause–effect structure, or MICES, to a flock of birds, and I find it a helpful analogy. A flock is irreducible, in the sense that no single bird can create one, and past structures determine future ones. Moreover, I can see how this structure is optimised to make a big difference to itself, because the moves of any single bird have a dramatic impact on the flock.[58]

The MICES could, in principle, be made of biological neurons or manmade computer chips. Yet the theorists are convinced that the current generation of computer chips would not allow the type of integration needed to generate a conscious structure. As Koch explained to me by Zoom in January 2025:

> There's nothing magical about brains that an engineered artifact couldn't, at least in principle, also possess. But the typical architecture of today's von Neumann machines, where you have one transistor talk to two or three or four other transistors and get input from a very, very small number of others, is radically different from the connectivity that a pyramidal cell has in your brain cortex. Some of these pyramidal cells get input from 20,000–30,000 cells and project to 20,000 or 30,000 other cells, all with massive overlap … It could be different if you built so-called neuromorphic computers, which are designed in the image of the brain – at the level of the silicon, they'll be built like a brain is built. These things could, in principle, be as conscious as you are.[59]

Tononi emphasises the MICES is equivalent to the experience of consciousness. It is "not correlated with … an experience, it is that experience". In other words because the MICES is targeted at explaining the phenomenon of consciousness, not its mechanism, it eliminates the explanatory gap.

As Koch put it, "How do you go from the mechanism to consciousness? ... That's sort of the hard problem. But IIT circumvents all of that; it starts out with consciousness."[60]

While IIT does not begin with circuitry, it does end with it. For Tononi, Koch and other IIT theorists, the brain circuitry that best meets the requirements of a MICES lies in a sensory region at the back of the brain where the neuronal circuitry has a grid-like architecture.[61] They've dubbed it the "hot zone". As Koch explains, this agrees with some experimental data that locates the moment of conscious perception to the back of the brain.[62] It also fits with evidence from people with localised brain damage. For instance, a patient who has suffered a stroke to the posterior cortex is able to see a keychain dangled in front of them but cannot identify it as a set of keys – a condition known as *agnosia* or absence of knowledge. Conversely, people who have up to a third of their frontal lobes removed show changed behaviours, but their consciousness remains intact, suggesting it must be located elsewhere in the brain.

Plumbing the depths of the unconscious brain

A person who is comatose or asleep clearly exhibits a different degree of consciousness to a person who is awake. IIT holds that the degree of consciousness is equal to the extent to which a system generates more information than the sum of its parts, a quantity Tononi has dubbed ϕ or *phi*. *Phi* can be measured in simulations, but the *phi* of the human brain is currently beyond measurement. However, in 2005, inspired by IIT and its prediction of integrated information, Tononi attempted to measure a crude proxy for *phi* in the human brain.

If the theory were correct, he reasoned, the brain of a clearly unconscious person, such as someone undergoing dream-free sleep, would have less integrated information than the brain of an awake, conscious person. This proxy for *phi* was measured using a variation of a traditional EEG. During dream-free sleep, the EEG shows a fairly regular, slow wave pattern, known as an alpha wave. Upon waking or dreaming, the EEG shifts to higher-frequency beta or gamma waves. Tononi and his colleague Marcello Massimini, a neurophysiologist at the University of Wisconsin–Madison, wondered if more information could be squeezed out of these EEG patterns. What if they were to zap the brain with a magnetic pulse? Like using sonar to plumb the structure of the ocean floor, perhaps the reverberations would fathom the structure of the unconscious brain?

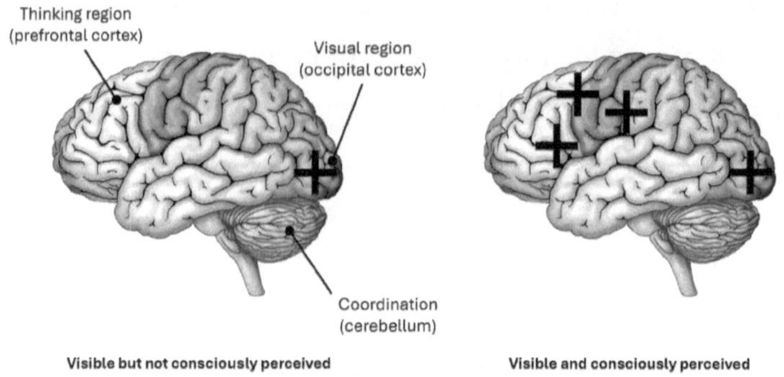

Figure 12. The signature of consciousness? When a person is not conscious of a clearly visible image (as on the right), the fMRI signal is restricted to the rear of the brain, around the visual region. When they report being conscious of the image, the signal moves to the front of the brain. But does that new signal reflect conscious perception or the act of reporting it?

They carried out a pretty cool experiment.[63] A transcranial magnetic stimulator – a metal wand with a double coil at its tip – was held above the subject's head to deliver the magnetic zap, rather like striking a bell. The reverberating electric pulses were recorded as an EEG from a cap studded with electrodes. The informational complexity in this EEG was analysed by "zipping" – a term familiar to anyone who has compressed a computer file. A simple file, such as a recording of a speech at a wedding, would zip to a low value. The more complex discordant chatter straight after the speech would zip to a high value. The measure of these values is called the perturbational complexity index (PCI).

It turns out that for a conscious person, or someone in the dream phase of sleep, the PCI is high; signals reverberate far and wide, as in a resonant bell. For a person in a vegetative state or in non-dream sleep, the PCI is low; signals barely travel. The higher the PCI, the more likely the patient is experiencing consciousness. Ongoing clinical trials with patients in an unresponsive vegetative state have been able to identify patients with a high PCI value who may well have residual consciousness, despite appearing unconscious from the outside.

While the PCI supports IIT, Anil Seth and Monash University philosopher Tim Bayne point out that it also supports other theories of consciousness that emphasise integrated circuitry, including the GNWT.[64]

Where the theories agree

IIT and GNWT (and many other theories of consciousness) agree that integrated circuitry is a core part of consciousness. Theorists from both camps cite the cerebellum as evidence for that. Tucked below the back of the brain, it looks so distinct that early anatomists dubbed it the "little brain". Its key role is to fine-tune body movements. People who suffer injuries to the cerebellum or who, in rare cases, are born without one, have trouble playing tennis and may have slurred speech, but they otherwise carry on normal lives. They are undeniably conscious. The cerebellum offers a compelling clue as to what it *doesn't* take to be conscious. It's not about the number of neurons; the cerebellum contains four times as many neurons as the human brain's "thinking" powerhouse, the neocortex. Nor is it about the density of connections between neurons; the individual cells of the cerebellum, known as Purkinje cells, have 200,000 connections, far more than the pyramidal cells of the neocortex, whose connections number in the tens of thousands.

What the neocortex has that the cerebellum doesn't is a wiring strategy that integrates information. The neurons of the cerebellum tend to transmit their information in one direction. That leads to the smooth information flow we need for automated body movements. By contrast, local feedback loops dominate the neurons of the cortex , so information reverberates in local eddies – like thoughts. IIT and GNWT agree that the one-directional circuitry of the cerebellum is unlikely to generate consciousness. Both agree that the neocortex, with its more highly integrated circuitry, is likely the home of consciousness.

Where they disagree

And now to the pointy end of things: where the two theories disagree.

For a start, they differ on what they consider consciousness to be. How can two theories compete if they don't even agree on what it is they are trying to explain? Apparently, they can. We'll come back to this later.

For GNWT, a core part of consciousness is cognition or "thinking". Consciousness arises in the workspace, which broadcasts a conscious perception across the brain. The thinking part of the brain, the prefrontal cortex, is a key component of this conscious workspace.

IIT does not consider thinking to be a core part of the conscious state. As Koch told me, "To think, to speak, to introspect, to rationalise, to justify, to do moral reasoning … those things may be related to, but they are different from, consciousness." For Koch, consciousness is a state of being. "You're angry; you're in love; you see red; you have a near death experience."

For Dehaene, "it's a big difference between our theories". "I don't believe in purified consciousness; the workspace is there for a function …

It's for distributing information so you can use it, memorize it, make decisions."[65]

It's an intriguing bifurcation that goes to the question of whether consciousness is about thinking, as in Descartes' "I think, therefore I am", or about "not thinking", as in the state experienced by a meditating yogi.

Beyond this major conceptual difference between the theories, they also make different predictions about how conscious experience plays out. The simplest difference relates to *where* it takes place. For GNWT, conscious perception involves a network of circuits across the brain, but crucially also includes the thinking part of the brain, the prefrontal cortex. IIT, by contrast, holds that consciousness involves *only* the rear sensory areas of the brain.

Another relatively straightforward difference concerns *timing*. Consciousness can be likened to a movie in which each fleeting frame is a conscious perception, such as the face of Scarlett Johansson suddenly popping up on a screen. Most consciousness experiments employ hacks to "freeze-frame" the movie, allowing scientists to isolate and study the properties of a single perception, such as when the subject first becomes aware of Scarlett Johansson.

According to GNWT, the conscious perception will be heralded by the sudden ignition of the workspace at the front of the brain, about a third of a second after the exposure to the new face. Moreover, the faster brain cameras used in the new experiment should also be able to detect an off signal, one that refreshes the workspace after the initial stimulus. (The predicted "off signal" was one of the fruits of the adversarial collaboration that challenged the adversaries to harvest new predictions from their models.) By contrast, according to IIT, the signal for each new conscious perception should persist uninterrupted for as long as it is visible.

Place and timing: they sound simple enough questions to resolve. You might wonder why this hasn't yet been nailed.

Here's the problem. Consciousness, as we've noted, is a poorly defined beast. Imagine bounty hunters setting off to capture a yeti. All they have to go on are some footprints and vague, differing descriptions from folk who claim to have seen one. The hunters lay their traps. They're all designed differently, based on each hunter's best guess as to the size, shape and habits of a yeti. They also place cameras at each trap site.

In come the results. One camera shows a glimpse of what looks to be an elephant with an oxpecker on its back (the oxpecker is a bird that pecks at the elephant's parasites). Another shows what looks to be a tail, but the rest of the creature is not visible. The hunters who trapped the elephant are criticised for making their trap too big; they have captured a bird unrelated to the hunt. The hunters who caught the tail are criticised for making their trap too small; the whole creature could not fit. And then there's the problem of different cameras. Some were too coarse-grained to give a clear picture of the beast; others were too slow to snap the beast before it ran off.

The yeti traps provide an analogy for the different psychological experiments used to try to capture the beast of consciousness. Experiments that require subjects to report being conscious of something may inadvertently trap beasts, like the oxpecker, that have a symbiotic relationship with consciousness but are considered separate from it.[66] These include mental processes such as attention, working memory and decision-making. Experiments that don't rely on subjects' own reports, such as binocular rivalry, may set the trap too narrowly and only capture a sliver of the target. Similarly, the type of "camera" used influences what you're likely to capture. A slow-speed camera like fMRI, which tracks blood flow to active regions of the brain, may be too sluggish to capture the signals of consciousness. EEG and magnetoencephalography

(MEG), which track electrical brain signals, are fast, but the trade-off is poor spatial resolution.

Another difference is that some studies look across multiple regions of the brain and average the results (multivariate), while others are more fine-grained, focusing on a specific part of the brain (univariate).

Arguments have raged for more than a decade as to whether or not these experimental traps are catching the right beast.[67]

The experiments

It took close to a year of wrangling for the adversaries and the referees to agree on the design of a new set of experimental traps to capture consciousness, no matter what form it might take.[68]

They would use three brain scanners: fMRI, MEG and intracranial EEG. This last, thanks to the altruism of thirty-four people with severe epilepsy. They agreed to participate in the consciousness experiments while undergoing tests to diagnose the source of their seizures. This involved having dozens of electrodes fitted to the surface of their exposed brain.

Using three different brain scanners requires three labs, each specialising in one of the different scans. And because each experiment had to be independently replicated, that would double the number to at least six labs. Each lab had to be "theory neutral", meaning they could not be affiliated with either of the key theorists, IIT's Giulio Tononi and GNWT's Stan Dehaene. A total of 256 experimental subjects were recruited.

Besides the types of scanners used, the experimenters employed machine-learning algorithms known as "decoders" to read the brain patterns. Over the last two decades these decoders have learned how to decode brain language by matching a particular pattern of brainwaves to a face, an object, an excerpt from a movie and even thoughts.[69]

Besides the technically dazzling tools at their disposal, the most important part of the experiment was the design of the consciousness trap. Different designs could lead to very different results. The adversaries were being asked to come up with something new, but they found it hard to agree. Ultimately they came up with a compromise. There would be two experiments. The first was preferred by the IIT camp, the second by the GNWT camp. The two experiments are described in detail in the following section. If it's too much, skip to the results in the final section, headed "Steeplechase".

The first experiment broke new ground. In particular, it attempted to pare away a bystander that had been getting caught in the consciousness trap: attention. Here, in simplified form, is what this pure consciousness trap looked like.

The subject viewed a computer screen that projected a sequence of up to thirty-eight clearly visible images while having their brain scanned. These images included faces, objects and letters. The subjects were conscious of them all, as ascertained by a surprise memory test straight after the main viewing. What varied was the amount of attention they paid to each of the images.

So how did the experimenters vary the amount of attention that the subjects paid to each image? The assumption is that if a subject finds something relevant, they will pay more attention to it. Before each viewing, the subjects were shown a picture that was nominated as the target, say the face of Scarlett Johansson. Then they were shown a sequence of other images. Pictures of other faces would appear "relevant" to them. By contrast, objects or letters would appear "irrelevant". So the subjects would be expected to pay more attention to the relevant images of faces than to the irrelevant objects or letters.

In the next experiment, the nominated target would swap to being the letter A, so during the viewing, all fonts would appear relevant, but faces or objects would be irrelevant. This generated two sets of data; one for *relevant* images; the other for *irrelevant* images. These data sets were used to train the brain decoder. First, it was trained on a relevant data set, say for relevant faces. So the decoder learned to identify the brain patterns associated with viewing relevant faces. Then the decoder was fed the brain patterns associated with the data set for irrelevant faces. Now it was up to the decoder. Once it had been trained to learn the brain patterns for faces when those faces were "relevant", could it still identify the brain pattern for faces when the faces were "irrelevant"? And if it found that there were common brain patterns for relevant and irrelevant faces, where in the brain did they lie?

In other words, the degree of attention that the subjects paid to the target varied. So too would the corresponding signals detected across the brain. But consciousness itself should not change. The signals which stayed the same when attention levels varied should capture the beast of consciousness.

The adversaries made three different predictions about the nature of the consciousness beast.

First, *where* it would be found:

GNWT predicted that the signature of consciousness would be seen in regions across the brain that included the prefrontal cortex.

IIT predicted that the common signature would only be seen in the posterior hot zone.

Second, *when*:

GNWT predicted that the "ignition signal" of a new conscious perception would enter the workspace at just less than a third of a second (around 300 milliseconds). Then activity would decay to baseline until an "off signal" marked the clearing of the perception.

IIT, on the other hand, predicted that the activity in the hot zone should persist as long as the duration of the visual stimulus, which varied between 0.5 and 1.5 seconds.

A third prediction related to the *connectivity* across the brain during a conscious percept. GNWT predicted that neurons across the frontal and rear parts of the brain would resonate in synchrony. IIT predicted that only neurons in the posterior hot zone and the earliest visual areas would resonate.

That's it for the first experiment. It is now complete, and its result provides the dénouement of this chapter.

A second experiment favoured by Dehaene is yet to be completed at the time of writing. In designing this consciousness trap, the key feature is to clearly demarcate when the subject is and is not conscious of a target, without requiring the subject to report seeing the target. "Reporting", as you'll recall, invites mental bystanders such as decision-making or short-term memory. The experiment is reminiscent of the "invisible gorilla". Participants play a video game something like Pac-Man, while in the background random images of faces and objects flash onto the screen. The most avid gamers will likely not be conscious of them. All the while, brain scanners are recording their brain patterns. Some of those brain patterns represent conscious awareness of the background images; others represent the pattern of a brain that is not conscious of them.

How do we know which is which? Every once in a while, the game stops and the gamers are asked if they saw the background images. Such retrospective reports eliminate the problem of mental bystanders while clearly demarcating which gamers were conscious of the background images. In this trap, the consciousness beast will correspond to a novel brain signature that is only seen in the gamers who reported seeing the background images.

Steeplechase

By May 2023, I'd spent many months researching the story of the consciousness contest for this book. The referees, Mudrik, Pitts and Melloni, were wonderfully generous about explaining the background and rules of the adversarial collaboration, but they were not permitted to share the results – which, by then, they knew! The results were to be presented on 23 June 2023 at the annual meeting of the Association for the Scientific Study of Consciousness (ASSC) at New York University, located near Washington Square in Greenwich Village. While there had been a lot of media attention at the start of this long-running contest, I was one of the few still paying close attention. I sniffed a scoop! I pitched the story to *Science* magazine, and they were in. So off I went to storied Greenwich Village, home of many a consciousness-raising experiment, to witness the showdown. It did not disappoint.

Figure 13. The author captured the moment when neuroscientist Christof Koch (standing) ceded the bet to philosopher David Chalmers: the correlates of consciousness had not been nailed. Greenwich Village, 13 June 2023, at the twenty-sixth meeting of the Association for the Scientific Study of Consciousness.

One the evening of 23 June, after a full day of mind-bending talks at NYU, the conference-goers walked a block to the university's Skirball

Center theatre. By 7.30 pm, the packed audience was buzzing like a Super Bowl crowd. The excitement was stoked by the anticipation of not just one showdown but two. Besides the result of the first adversarial collaboration for theories of consciousness, it would also see the calling of the 25-year bet between two legends of the field, David Chalmers and Christof Koch.

Before these grand finales, the audience was treated to the warm-up acts.[70]

First, a welcome from NYU philosopher Ned Block, co-host of the conference with David Chalmers. As MC, Heather Berlin, a clinical psychologist, neuroscientist and documentary maker, set the scene with the notion of "creative destruction" and how "science advances through the graveyards of theories".

All the big names were there: Daniel Kahneman, seated in the second row just in front of me; Daniel Dennett, in poor health, addressed the audience on video; and of course, the adversaries and neutral participants of COGITATE – the name given to this adversarial collaboration.

To give the backstory of the bet, a Swedish journalist, Per Snaprud, recounted how he'd met Chalmers at the gathering of consciousness researchers held twenty-five years before in Bremen. Chalmers had spent the night philosophising with Koch in smoky bars. Snaprud encountered him the following morning and heard about the wager that had been made just hours before: that the neural correlates of consciousness would be nailed in the next twenty-five years. Sniffing a big story, Snaprud recorded the details on his tape recorder, noting: "Chalmers wears a black leather jacket and had shoulder-length hair ... even though he hasn't slept, his voice is remarkably alert." It was lucky that Snaprud made the recording, because Chalmers and Koch had forgotten the exact terms of their bet. Snaprud was able to find the old cassette and provide them.[71] It turned out they were pretty vague, merely describing "a small set of neurons characterized by a small list of intrinsic properties [less than ten], and it will be clear in twenty-five years." Koch bet in favour of the prediction, Chalmers against.

Next, the theorists Melanie Boly, standing in for Giulio Tononi and IIT, and Stan Dehaene, representing GNWT, gave brief overviews of their theories. Neutral referee Liad Mudrik described how COGITATE had designed the experiment to tease apart the predictions made by the two theories.

Finally, the baying crowd got what they came for. Lucia Melloni took to the stage for the big reveal. Based at the Max Planck Institute, she began by reminding folks of Planck's motto: "science advances one funeral at a time". "Perhaps," she said, "the adversarial collaboration was a way not to have to do that." Then she launched in. Writ large on a screen above her were the predictions that had been made based on each theory, and, on the other side of the screen, the results.

It reminded me of a steeplechase: the two theories were ponies, and the predictions were hurdles. When a theory cleared a hurdle, it scored a green light; when it did not, the light was red.

The first hurdle tested which parts of the brain were most strongly associated with conscious perception. There were two parts to this hurdle: perceiving the image and perceiving its orientation.

IIT scored two green lights, because the rear areas of the brain performed well on both tasks. GNWT scored a single green light; the front of the brain was good at perceiving images but not their orientation. So, on the first hurdle, IIT appeared to pull ahead.

The second hurdle tested the timing of the signals. IIT predicted that the signal in the posterior "hot zone" would last for as long as the stimulus did, which in this case was 0.5, 1.0 or 1.5 seconds. IIT scored a green light for this one.

GNWT predicted an ignition followed by an off signal. The ignition signal was detected, at around 300 milliseconds, but not the off signal. GNWT failed its full prediction, scoring a red light.

So at this second hurdle, IIT pulled further ahead.

The third hurdle measured overall connectivity. Which regions of the brain were "synchronised" during conscious perception? IIT predicted there would be sustained synchrony between the earliest visual processing areas and the posterior hot zone. But here the IIT pony faltered; no sustained synchrony was observed. IIT scored a red light.

GNWT, meanwhile, predicted that while the workspace was active, there would be a brief period of sustained synchrony between the frontal regions and the rear sensory regions. The GNWT partially cleared this hurdle, scoring a partially green and partially red light.

With that, the score was tallied. Overall, IIT had more green lights.

...

It felt like a winner had been called. Melloni did her best to discourage that impression. "I want to emphasise that this is just one piece of a puzzle," she told the audience. Importantly, the second part of the adversarial collaboration, the experiment preferred by the GNWT camp, was yet to be completed.

Melloni shared that early in the process, she'd asked Kahneman, the "father of adversarial collaborations", to come and talk to the COGITATE group. He had told them: "People don't change their minds." "I hope you will prove him wrong," Melloni said.

With that, Melanie Boly of the IIT team and Stan Dehaene of the GNWT team took to the stage. They sat awkwardly next to each other. Boly wore a triumphant expression. Dehaene's demeanour was stoic.

Asked for her response, Boly declared, "The results corroborate

IIT's overall claim that posterior cortical areas are sufficient for consciousness, and neither the involvement of [the prefrontal cortex] nor global broadcasting are necessary." As to the stumbles at some of its hurdles, Boly pointed to technical limitations such as a lower density of recording electrodes in the rear of the brain during intracranial EEG.

When it was Dehaene's turn to respond to the results, he quipped: "I've decided to follow the advice of Dan Kahneman." In other words, he had not changed his mind. Overall, he said he was pleased. "Many aspects of the result fit with the theory ... the most interesting part of this experiment was the task-irrelevant stimuli." The question was whether they would indicate the ignition of a conscious perception in the frontal brain. "The answer is yes!" he said. As to the failure of GNWT to clear some of its hurdles, Dehaene suggested that the experimental design may have allowed the subjects' minds to wander. It was, after all, not his preferred design.

More cuttingly, he added that the hurdles were set lower for IIT than for his GNWT. "None of the mathematical backbone of the IIT is being tested here; none of the phi; none of the axioms; none of that." And he repeated something that philosopher Ned Block had pointed out in his opening remarks: the results for IIT would fit many back-of-the-brain theories including Block's preferred First Order Theory. "I think it's a bit superficial, just getting the right colours," Dehaene grumbled, referring to IIT's win based on green lights. The adversaries had lived up to Kahneman's maxim.

Calling the bet

But the evening was not over. There was a bet to call.

Chalmers and Koch ascended to the stage, the philosopher clad in his signature black and the neuroscientist in his signature psychedelic hues.

"It's not about the bet", began Chalmers, eliciting uproarious laughter from the crowd. "The results are amazing; it's an incredible achievement … Arguably GNWT was challenged more seriously than IIT, but there were still serious challenges for both. We're still in the middle of the scientific process, which is a good place to be for science, also a good place for me … At the very least, it's clear that things are not clear," he philosophised.

Then it was Koch's turn. "I'm going to concede that we have not achieved convergence in the general community about the neural correlates of consciousness."

After rummaging around backstage for a moment, Koch gallantly handed over a bottle of 1978 Madeira, with five more fine wines in the crate.

"I've lost the battle," he declared, "but won the war for science."

…

I got my scoop! I reported this story for *Science* magazine on 25 June, barely a day later.[72] I reported what I experienced – an event that had felt like a sporting contest, with a winner declared, albeit with qualifications. I copped some virulent criticism for reporting it this way.

Eighteen months on, I went back to take another look at the results of COGITATE. The day of the Skirball event, their preprint had gone online describing the results.[73] It had been a tough read at the time; I struggled to understand the convoluted design of the experiment and its interpretation. But this time, without the nerve-racking deadline, I read my way through it very slowly, and with help from Liad Mudrik and Michael Pitts, who were endlessly patient in answering my questions.

I found myself in awe of the vast amount of data acquired by the

COGITATE team. It had been meticulously analysed. Crucially, they had not crowned a winner or loser but laid bare the strengths and weaknesses of each theory.

As Melloni, Mudrik and Pitts wrote:

> Beyond challenging the theories themselves, we present an alternative approach to advance cognitive neuroscience through a principled, theory-driven, collaborative effort. We highlight the challenges to change people's mind and the need for a quantitative framework integrating evidence for systematic theory testing and building.[74]

So where does this leave the science of consciousness? Can normal science – the sort that, according to Kuhn, sees scientists doggedly seek to fortify their own theories – be kickstarted by adversarial collaboration into a scientific revolution? There have been mixed reviews. In the immediate aftermath of the Skirball event, the critiques were strident. Some, such as computational neuroscientist Megan Peters at the University of California, Irvine, bristled at media coverage (like mine) that reported the results of the adversarial collaboration as a two-horse race. "It's important to see that science advances by learning from each experimental hurdle," she told me. But Peters remains a fan of adversarial collaborations. During Covid lockdowns, she was inspired to co-organise a series of debates under the banner of Generative Adversarial Collaborations.[75] "Watching the teams chew on stuff was instructive," she told me.

Olivia Carter, a neuroscientist at the University of Melbourne and past president of the Association for the Scientific Study of Consciousness, believes the two theories were too far apart for their predictions to be meaningfully compared. "My personal feeling is that they are testing totally different things," she said.[76]

It's an apt criticism. As noted above, the two theories do indeed have different ideas of what consciousness is. Even so, Liad Mudrik points out that the theorists all agreed that the experiment they co-designed targeted what they considered to be visual consciousness.

Monash University philosopher Jakob Hohwy, a participant in another Templeton tournament, agrees. He points out that the field is still divided over such fundamentals as the definition of consciousness and whether it is closer to thinking or feeling. For Hohwy, "We will find out as we go along in exactly this type of adversarial collaboration."[77]

That's Anil Seth's view, too. (He is participating in the adversarial collaboration with Hohwy.) The "extremely valuable [findings] … will push forward the development of both IIT and GNWT and other theories of consciousness, too, by providing new constraints and new explanatory targets."[78]

Overall, it seems impossible to fault the lofty ambition of this first adversarial collaboration for the science of consciousness. It broke new ground, forcing theorists to make more tangible predictions, and it made experimentalists work out new techniques. It committed to transparency and integrity, and the experiments were carried out to a Rolls Royce standard. The results are now available for all to pore over.

For Melloni, the fact that the adversaries have not changed their minds does not detract from the value of the process. "As Kahneman says, people don't change their mind, yet the way they react to the challenges makes their theory progress or degenerate," she said. "If the latter, over time the theory 'dies' and scientists abandon it."[79]

> In Kahneman's words: The path on which a theory is set eventually becomes obvious to everyone – except perhaps the theorist. And this is how adversarial collaborations can advance science without adversaries changing their mind.[80]

And what of Christof Koch? He'd offered Arthur Eddington's disproof of Newton's theory during the 1919 eclipse as the exemplary "adversarial collaboration". But the adversarial collaboration for consciousness did not disprove any theory. In 2023, Koch argued, the goalposts are different:

> Physics deals with very simple systems. Here you had one beam of light, you know, close to one star, vastly simpler than something like theories of consciousness ... With physics you make an assumption, either it holds or doesn't hold. In neuroscience, you have to make all sorts of auxiliary assumptions. And so, some of these theories could probably be rescued, but in a different form.[81]

Yet even Eddington's eclipse observations didn't deliver an overnight scientific revolution. Eddington was criticised for cherry-picking the data from his telescope images, which were compromised by cloudy conditions and lenses warped by the hot humid atmosphere.[82]

It took many more eclipse expeditions to confirm Einstein's theory of General Relativity. A century later, we're still testing Einstein's theory.

I guess we'll be doing the same with theories of consciousness.

Postscript

Three months after the report on the results of the adversarial collaboration at the conference in Greenwich Village, a major ruckus broke out in the consciousness community. The first blow was a letter, posted as a preprint, that accused IIT of being a pseudoscience. It was signed by 124 highly respected researchers from the neuroscience and philosophical communities.[83] A year and a half later, in March 2025, a final version with 103 authors was published in *Nature Neuroscience*.[84]

The accusation centred on two things. First, that while IIT put some of its predictions to the test, its core prediction – that the conscious state was equivalent to integrated information (phi) – was unmeasurable and hence untestable. Second, that the champions of IIT misrepresented the theory. Notwithstanding the untestability of its core claim, they promoted it as a leading theory. Moreover, the theory lent credibility to "panpsychism", the idea that "the universe is suffused with sentience", which had resulted in its practitioners being celebrated by the pseudoscience community.[85]

Neuroscientists Alex Gómez-Marín at the Institute of Neuroscience in Alicante, Spain, and Anil Seth at the University of Sussex argue eloquently against these charges.[86] They cite other theories that have failed to provide experimental tests yet offer a valuable conceptual framework for scientists, such as string theory. And, poignantly, they highlight another esoteric theory that was subjected to "trial by authority": Einstein's theory of General Relativity. In 1931, 121 academics signed a series of essays published as a book titled *A Hundred Authors Against Einstein*, rejecting his theory as "as a fundamentally misguided and logically untenable fiction" and complaining that it was being given undue publicity.[87] Gómez-Marín and Seth conclude: "as long there are productive iterations of theory development and experimental tests, we are in the business of science".

5

Gravitational Waves

The grand aim of all science is to cover the greatest number of empirical facts by logical deduction from the smallest number of hypotheses or axioms.

Albert Einstein[1]

I ought to know by this time that when a fact appears opposed to a long train of deductions it invariably proves to be capable of bearing some other interpretation.

Sherlock Holmes[2]

In my scientific career, I studied a gene that designed the fourteen segments of a fruit fly. Each of the segments, numbered 1–14, had different features. To learn more about the gene, I chopped it into pieces and made a surprising discovery. One part of the gene carried the instructions for building the even-numbered segments of the fruit fly; the other the odd-numbered segments.[3] It allowed me and my colleagues to induce a theory about how the gene works. I thought all scientists operated this way: gathering data and then inducing theories based on that data.

So I was shocked to discover that one of the most consequential theories ever developed, Einstein's theory of gravity, emerged through an entirely different process. General Relativity, as the theory is formally known, was not induced from data. Einstein did not do any experiments. He developed his theory from first principles via the process of deduction. Among its key predictions were the existence of black holes and gravitational waves. For over a century, Einstein's theory has reigned supreme, repelling all efforts to topple it from its throne.

Telling the story of how it has done so makes for a riveting lens into the twentieth century.

Its *dramatis personae* are Promethean characters. Einstein was driven to "know God's thoughts".[4] John Wheeler, coiner of the term black hole, exhorted physicists to "find the strangest thing and then explore it".[5]

It is a tale of the power of science. If the Nazis derided Einstein's theory as "Jewish science", the Americans used $E=MC^2$ to build a bomb to fight the Nazis. The exodus of German scientists supercharged American science. Prior to 1930, Germany won the lion's share of Nobel Prizes; after the 1930s, that title went to America.

It is a tale of scientific sin. After Hiroshima and Nagasaki, Einstein was mortified by what his physics had unleashed. He threw himself into the pacifist cause, as did "the father of the atomic bomb", Robert Oppenheimer – whose words after witnessing the mushroom cloud in the Los Alamos desert were: "Now I am become Death, the Destroyer of Worlds."

Finally, it is a tale of turning weapons to ploughshares. Atomic bomb physicists used their newfound knowledge of atomic forces to predict the fate of stars. After building a hydrogen bomb equal to 1300 Hiroshimas during the Cold War to keep the Russians at bay, Wheeler became the "father of modern general relativity" by returning to the lab to test Einstein's predictions of black holes and gravitational waves.

The twenty-first century would once again prove Einstein's theory correct.

...

Physicists are engaged in the modest quest to unlock the secrets of the universe.

They also believe, based on experience, that those secrets are revealed by solving paradoxes. "How wonderful that we have met with a paradox," said Danish quantum physicist Niels Bohr. "Now we have some hope of making progress."

The simple idea of relativity – that all motion is relative to the perspective of the observer – has long provided fruitful paradoxes.

Let's start with the solar system. Ptolemy, in the second century CE, placed the Earth at the centre, with the sun and the planets going around it. The problem was, every so often the planets appeared to reverse their direction of travel in the sky. Astronomers needed complicated adjustments to make Ptolemy's model work, such as proposing that planets performed an impromptu backwards loop at one point of their orbit. It wasn't until 1514 that the Polish priest Nicolaus Copernicus took a different perspective and solved the paradox. If the Earth and the planets actually moved around the sun, then the reversals were just an illusion. As a faster-moving planet like Earth approaches and passes a slower one like Mars, the slower one will for a while appear to be travelling backwards. It's like passing a car on a highway. To you, the slower car appears to be moving backwards.

A hundred years later, in 1613, Galileo Galilei backed up Copernicus' explanation with his observations of Venus. The planets were indeed all orbiting around the sun. So, solving the paradox of reversing planets revealed the true configuration of the solar system.

But why did the planets orbit the sun? Gravity, proposed Newton in 1687, the same force that causes the apple to drop from the tree. He did not, however, attempt to explain *how* gravity could exert its force over such vast distances. "*Hypotheses non fingo,*" he said: "I make no hypotheses."

In 1905, Einstein came along. He was very much in the tradition of believing that paradoxes conceal the secrets of the universe. At this time, he was obsessed by one: a light beam does not obey the laws of relativity. In solving this paradox, Einstein would crack open a bizarre secret of the universe. Space and time are not the inert stage upon which the objects of the universe perform. Space and time are actors. Furthermore, they weave together to create the elastic fabric of the universe – spacetime. In conceiving of spacetime, Einstein would also come up with a theory of gravity. Gravity is a result of the elastic properties of spacetime. Like a bowling ball on a trampoline, massive objects, such as the sun, create a well in the fabric of spacetime, and smaller objects, such as Earth, find themselves pulled into that gravity well.

That is the gist of Einstein's theory of General Relativity. Its predictions, such as the bending of starlight by the sun during an eclipse, have been proven repeatedly over the last century.

But there was one prediction that held out: the existence of gravitational waves. These waves would carry the force of gravity, akin to the way electromagnetic waves carry electrical and magnetic force. It would take a century to prove their existence, working against the resistance of many scientists who believed either that they did not exist or that if they did, they were too faint to ever detect. One of the biggest sceptics was Einstein himself.

Origins of General Relativity

To tell the story of gravitational waves, we must begin with Einstein's theory of General Relativity.

Fear not. Even though the theory is underpinned by some diabolical maths, the concepts are surprisingly accessible. Einstein did not dive into equations to pluck out the pearl of his theory. He was a conceptual thinker renowned for his metaphorical thought experiments.

Like Sherlock Holmes, he deduced his way from one clue to the next to arrive at his final conclusions. What set him apart was his determination to go boldly where none had gone before, in his quest to "know God's thoughts".

It is well worth trying to follow the deductive stepping stones on Einstein's trail of discovery. Three Nobel Prize winners in physics described General Relativity as the "greatest feat of human thinking about nature", "probably the greatest scientific discovery ever made" and "one of the greatest achievements of human thought".[6]

Einstein's journey, like that of Copernicus, was triggered by a paradox involving "relativity". Let's bring it back to a familiar example. You're sitting on a train and notice a carriage on a parallel track pulling away from you. For a disorienting moment, you have no idea which carriage is actually moving. Indeed, relative to each other, your speed is the same.

Copernicus used the idea of relativity to explain the paradox of planets appearing to move backwards. Einstein, centuries later, was drawn to the paradox of the speed of light. This paradox was first uncovered in 1865 by Scottish mathematician James Clerk Maxwell, whose portrait, along with that of Newton, hung in Einstein's study. Maxwell had discovered that a light beam did not obey the principle of relativity.

Maxwell's discovery emerged from his attempt to discover the relationship between electricity and magnetism. That such a tryst existed was a surprise for the folk of the early nineteenth century. The two phenomena appeared quite distinct. Electricity could power a puny lightbulb, but it was mostly a party trick.[7] You could store up charge in a Leyden jar, ask an unsuspecting person to touch it and then watch their hair stand on end. Magnetism, on the other hand, was a fantastically useful property of certain metals that ensured a needle on a compass would always point north. There was no reason to suspect a link between the two. Until, in 1820, Danish physicist Hans Ørsted happened to have a compass on his work bench when he connected a wire to a battery. As the current gushed forth, he was astonished to see the needle jump. Conversely, in 1831, Michael Faraday found that by sliding a magnet through a coiled wire, he could trigger a current to flow.

So, an electric current could move a magnet, and a moving magnet could trigger an electric current. But what exactly was the mathematical relationship between electricity and magnetism? Maxwell spent over ten years trying to discover it. It turned out not to be a straight line or a curve, but a wave. In his 1865 paper, Maxwell's equations showed that this wave had a speed of 300,000 kilometres per second.[8] That was a very familiar speed: it was the speed of light, which had been measured by various means since the late seventeenth century.[9]

Maxwell's equations were a revelation. They showed that the interaction between electricity and magnetism was something akin to the way a Mexican wave moves through a sports stadium. One group stands up then sits down. That triggers the next group to do the same. Likewise, the electric impulse triggers the magnetic impulse, which triggers the electric impulse and so on. We detect those mutually perpetuating impulses as light.

All of this is familiar to us now. But in 1865, the fact that electricity, magnetism and light were all part and parcel of the same thing was the revelation of the century. And it raised a conundrum. According to Maxwell's equations, the speed of the light wave was a constant 300,000 kilometres per second. But how could that be? Every speed is relative.

As a 26-year-old junior patent clerk, Einstein found this conundrum deeply troubling. In his bones, he felt it concealed a secret about the universe.

There was one ready answer: that speed was relative to *ether*, an imaginary medium through which light waves travelled in the same way that ocean waves travel through water. Einstein wasn't happy with that answer. For one thing, there was no evidence for the ether's existence. For another, this answer just didn't satisfy him – he sniffed a bigger mystery lurking. In his thought experiments, he obsessed about trying to catch up to a light beam in the way that a speedy bicyclist would gain on a lumbering rail car. But Maxwell's equations held that no matter the bicyclist's speed, the relative speed of the light wave would never change.

That was intolerable to Einstein. So, he took a step further and wrecked the foundations of physics.

The Scottish Enlightenment philosopher David Hume may have been responsible. Einstein had studied Hume as part of his philosopher's club in Bern, Switzerland. The club of three included a maths colleague from his university days, Conrad Habicht, and the dapper Maurice Solovine, a philosophy student who had come to Einstein for physics lessons. Over supper and cigars, the trio read works of free thinkers, such as *Antigone* by Sophocles and *Don Quixote* by Miguel de Cervantes. They paid particular attention to David Hume's 1739 *A Treatise of Human Nature*, which dissected the practice of empiricism – how we test theories with evidence. As we saw in the introduction, Hume warned about the dangers of extrapolating from observation to general laws of the universe.

Hume's famous example is the assumption that the sun will rise tomorrow. The sun has so far risen every day that humans have been observing it – but must it? Just because it has risen every day in the recorded past does not mean that it must rise tomorrow. Some five billion years hence, the sun will have expanded into a red giant and vaporised us. So there will be a tomorrow when the sun does *not* rise over Earth.

Einstein must have had Hume in mind when he took his leap.

If he could never catch up with a light wave the way a bicyclist can catch up to rail car, what could that mean? Speed is the time it takes to cover a unit of distance; say, how many seconds it takes to cover each metre. The assumption is that a second always lasts a second, and a metre always measures a metre. But if the universe dictates that you can never gain on a light beam, no matter how fast you race it, then for you, the racer, a metre is no longer a metre and a second is no longer a second. As you speed up, your metre starts to shrink and your second grows longer. In other words, it takes longer to cover a smaller distance. It sounds like the stuff of nightmares. But it explains why you can never catch up to that light beam.

So, is this crazy explanation true? Yes. It was proven so in 1971, when scientists Joseph Hafele and Richard Keating took atomic clocks for a round-the-world flight. Compared to stationary clocks on earth, the flying clocks slowed down. For the fast-flying clocks, the duration of each second had indeed lengthened. Satellites have to take this relativistic information into account; otherwise, Google Maps would send us all driving into ditches. One of the other things delivered by Special Relativity was the world's most famous equation: $E=MC^2$. With its revelation of the titanic energy locked up in matter, it opened a gateway to the development of the atomic bomb.

So that's Special Relativity. In trying to explain why a light wave moves at constant speed relative to everything else, Einstein had wrested a new secret from the universe. Space and time are not the fixed things we thought them to be.

...

Einstein published his theory in 1905 – but soon realised that it was far from complete.

Special Relativity described how things behaved when they were moving at constant speed, like a comet out in empty space. But once the comet's trajectory takes it past a planet, it starts accelerating. So, in the case of acceleration, what happens to space and time?

In 1907, Einstein benefited from another thought experiment. He imagined being in an elevator. As the elevator accelerates upwards, he suddenly feels extra heavy, as if the earth's gravity had increased. If he were standing on a set of scales, they would indeed show an increase. But now the elevator cable breaks, and he starts falling. He feels weightless, as though gravity has ceased to exist. Indeed, the scales taped to his feet register zero.

Einstein had what he calls "his happiest thought". He realised acceleration and gravity were indistinguishable.

In the second part of this thought experiment, Einstein imagined himself in an elevator falling in space under the influence of gravity. Then he takes out a torch and flashes it at the opposite wall, directly in line with his head. To him, the beam of light would appear to move in a straight line, illuminating a spot on the opposite wall. But to an outside observer, the beam of light would appear to fall, like a curving waterfall, to reach the same spot on the wall while the accelerating elevator fell many metres below the starting position. In other words, under the influence of acceleration, light curves. And since gravity and acceleration are indistinguishable, under the influence of gravity, light would curve. Ergo, spacetime itself curves under the influence of gravity.

OK, this thought experiment is tough to follow. I am reminded of what Chaim Weizmann said after accompanying Einstein on a 1921 Atlantic crossing from Europe to America: "Einstein explained his theory to me every day, and on my arrival I was fully convinced that he really understands it."[10] So let's not worry too much if we can't keep up with Einstein here. Bottom line: Einstein deduced that gravity causes the fabric of spacetime to curve.

The trampoline analogy of spacetime helps. Heave a bowling ball onto that trampoline and it creates a big well. Next throw on a tennis ball, which will fall into the well and roll around. The bowling ball is like the sun; our Earth is like the tennis ball rolling around it. According to General Relativity, that's how we should imagine gravity – not as a force but as a well in the fabric of spacetime.

Einstein had deduced his way from the paradox of an uncatchable light beam to a theory of gravity. And that led him, unintentionally, to take up where Newton left off. When Newton crystallised his laws of motion and gravity, he put concerns of relativity aside.[11] His laws worked fantastically well at accounting for the speed and shapes of planetary orbits, as well for the trajectories of cannonballs. And they set the gold standard for a scientific theory by making testable predictions, such as the timing of an eclipse over England in 1715 and the return of Halley's comet in 1759, after Newton's death. But when it came to actually explaining what gravity is, Newton didn't try.

That was left to Einstein, when, 228 years later, he presented his theory of gravity in a lecture to the Prussian Academy of Sciences on 25 November 1915. In that lecture, he chalked up his field equations – the mathematics that showed exactly how a mass would warp the fabric of spacetime. Another crazy-sounding theory? Einstein could prove it. Or rather, he needed astronomers to.

Just as Galileo had peered out at the heavens to prove that first theory of relativity – that the sun, not the earth, is the centre of the solar system – Einstein needed astronomers to prove *his* relativity theory. Specifically, by pointing their telescopes at the stars during an eclipse of the sun. Normally we only see stars at night, when they are well away from the sun's glare. But during an eclipse, the sun is blotted out for a few minutes by the moon and any stars lying alongside the sun are briefly visible. Einstein predicted that the light from these stars would curve along the gravity well created by the sun. Viewed from Earth, the image of the star cluster would be warped, as if by a lens.

Several astronomers mounted expeditions to try to photograph this warped image of a star cluster during an eclipse. But the First World War got in the way. (Einstein was lucky. His prewar maths included an error that meant his calculations were out; had the 1914 expeditions succeeded, he would have been proven wrong![12]) In 1919, Arthur Eddington, head of the Cambridge Observatory, was the first to succeed in testing Einstein's theory. His measurements showed that Einstein was right. "Lights all askew in the heavens", proclaimed a *New York Times* headline on 10 November 1919.

When Einstein learned of Eddington's results, a student asked him what he would have done if they had proven him wrong. "Then I would have to pity the dear Lord," he replied. "The theory is correct anyway."[13]

A key change

If this story were a symphony, Eddington's vindication of Einstein's theory would represent the triumphal climax of the movement. But the next movement is in a totally different key. Less Beethoven's *Eroica*, more atonal Schönberg. Because while Einstein had no doubt about the warping of starlight, when it came to the other predictions of his theory – gravitational waves and black holes – he did his best to discredit them.

Today, no one bats an eyelid at the notion of gravitational waves or black holes. But until the 1960s, most physicists did not take either concept seriously. They were the unloved offspring of Einstein's theory of General Relativity. Einstein's 1915 equations for General Relativity had predicted the existence of gravitational waves, largely by analogy, much as Maxwell's equations for electromagnetism had predicted electromagnetic waves. Yet there were such big differences between an electromagnetic field and a gravitational field that it made Einstein doubt the analogy. So much so that after first proposing the existence of gravitational waves in 1916, Einstein decided they didn't exist. Then he changed his mind. He went back and forth six times. His last paper on the issue, in 1936, written with his student Nathan Rosen, started off as a final attempt to refute their existence. "I arrived at the interesting result that gravitational waves do not exist," wrote Einstein to his colleague Max Born.[14]

Einstein and Rosen submitted their paper to the prestigious journal *Physical Review*. The reviewer, Howard Robertson, rejected it based on a mathematical error. Einstein was offended and submitted the paper to the lower-profile *Journal of the Franklin Society*, which did not require peer review. Just in the nick of time, Einstein realised Robertson was right and rewrote his paper with the opposite conclusion: gravitational waves could exist! Robertson's "friendly help" was acknowledged in the publication. Yet he still did not believe they could be detected.[15]

By the time Einstein died in 1955, no one had taken much interest in gravitational waves in a couple of decades. Yet by the 1960s, they were back on the radar. Joseph Weber, a submarine commander turned physicist, was building a "resonant bar" to detect their vibrations at the University of Maryland. Meanwhile, theoretical physicist Kip Thorne, at Caltech, was predicting what these gravitational waves might look like as they emanated from colliding black holes – notwithstanding that there was not a skerrick of evidence that they existed.

Today, we know that black holes exist. In 2022, the Event Horizon telescope showed us one at the centre of our galaxy. But in the 1960s, Thorne's belief in black holes was all theoretical. They had emerged from the musings of an astrophysicist ducking artillery shells. In 1916, Karl Schwarzschild, a German Jew, was serving at the Russian front, even though, at the age of forty-two, he was a little old for military service. Notwithstanding battle, the mail service was good. He received a copy of Einstein's paper on General Relativity a month after its publication, and began applying Einstein's equations to see what would happen to spacetime surrounding an object with extreme density. For instance, if you were to begin with an object with the mass of the sun and compress its volume, what would happen to spacetime? As the mass shrinks, the curvature of spacetime around it would increase. Once the sun compressed to a diameter of 3 kilometres – a threshold we now refer to as the "event horizon" – the surrounding spacetime would twist to the point that no object or light could ever emerge. It would become a black hole. Schwarzschild penned his calculations to Einstein, with a note saying: "The war treated me kind enough, in spite of the heavy gunfire, to allow me to get away from it all and take this walk in the land of your ideas."[16]

Tragically, he died some six months later, probably of an autoimmune illness.

Einstein liked Schwarzschild's maths but not his notion of a black hole. He considered it a *reductio ad absurdum* of his theory. Just as he had done with gravitational waves, in 1939 Einstein published a paper to prove why black holes could not exist.[17]

So how is it that by the 1960s, Thorne was proposing that gravitational waves might be detected from the collisions of black holes, when there was no evidence for the existence of either? It seems as fanciful as proposing an experiment to detect levitation in yetis. Physicist Federico Ferrini likened it to trying to detect "the flapping of an angel's wings".[18]

The blame for this fanciful proposition can be laid almost entirely at the feet of John Wheeler. Known for his wild ideas, he was eager to go boldly where others feared to tread. And he attracted like-minded souls. He reminds me of Ernest Shackleton and his expedition to cross the *terra incognita* of the Antarctic. Shackleton's advertisement to recruit men supposedly read: "Men wanted for hazardous journey, small wages, bitter cold, long months of complete darkness, constant danger, safe return doubtful, honour and recognition in case of success." Apparently

5000 people answered it. Like Shackleton, it seems, Wheeler had no shortage of recruits.[19]

Wheeler had spent the Second World War working for the US military. He'd built the plutonium bomb that destroyed Nagasaki in 1945. Then he had helped to build the H-bomb, the destructive power of which was equal to thousands of Nagasakis. By the late 1950s, he had returned to his true vocation: unlocking the mysteries of the universe. Chief among them was the fate of big stars. It was no accident that one of the fathers of the H-bomb would find himself drawn to the fate of large stars. That fate was determined by a yin-yang contest. On one hand, the crushing force of gravity; on the other, the outward push of atomic forces. Calculating the strength of the atomic forces was Wheeler's specialty. As Kip Thorne put it, "He was among the world's greatest experts on the laws of physics that govern the structure of matter."[20] It was those calculations that led Wheeler to the conclusion that black holes *must* exist.

A short history of black holes and gravitational waves

Oppenheimer was not the first to imagine a black hole. There was Schwarzschild in 1916. But even before him, John Michell in 1783 and Pierre-Simon Laplace in 1796 had independently conceived that particles of light would not escape from a star if its gravity was great enough.

Einstein was not the first to imagine gravitational waves. In the early 1800s, French philosopher and mathematician Pierre Laplace proposed gravitational waves as a kind of emanation from the sun that would create friction with the Earth and cause the Earth to spiral towards the sun. But Laplace did not have the right mathematical model and gave up on his theory.[21]

In 1893, self-taught English physicist Oliver Heaviside also proposed gravitational waves by analogy with Maxwell's discovery of electromagnetic waves. In 1905, French mathematician Henri Poincaré – to whom Einstein owed much for sparking his ideas on special relativity – also published a theory that gravity was propagated through a wave by analogy with Maxwell's equations.

Probing the fate of stars

When it came to pondering the fate of stars, Wheeler was partaking in a time-honoured tradition. As Shakespeare's Julius Caesar proclaimed:

> ... I am constant as the northern star,
> Of whose true-fixed and resting quality
> There is no fellow in the firmament.

Shakespeare's play appeared in 1599, so for over two hundred years he was on safe ground regarding the constancy of stars. It wasn't until 1844 that German astronomer Friedrich Bessel predicted that the brightest star in the sky, Sirius, must have a partner – one whose story would reveal that the lives of stars were spectacularly *inconstant*.

Bessel's observations showed something invisible tugging on Sirius as it moved brightly through the night sky. Eighteen years later, in 1862, US telescope maker Alvan Clark found Sirius B – the first white dwarf ever to be discovered. Sirius B was not only 10,000 times dimmer than Sirius; calculations showed that while its mass was similar to that of our sun, it was packed into a volume the size of Earth. That gave it the density of around one ton per cubic inch, 3000 times denser than anything on our planet.[22]

The white dwarf must have been composed of an exotic new form of matter. In his signature prose, Arthur Eddington expressed the general sense of befuddlement:

> We learn about the stars by receiving and interpreting the messages which their light brings to us. The message of the companion of Sirius when it was decoded ran: "I am composed of material 3000 times denser than anything you have ever come across; a ton of my material would be a little nugget that you could put in a matchbox." What reply can one make to such a message? The reply which most of us made in 1914 was – "Shut up. Don't talk nonsense."[23]

Nevertheless, an explanation was required. Eddington offered a partial one in his 1926 textbook *The Internal Constitution of the Stars*. Sirius burned bright because of hydrogen fusion, an alchemy that created one helium atom from two hydrogen atoms and released fantastic amounts of energy. Sirius B had exhausted its store of hydrogen, which explained its pallor. And its diminutiveness? Without the outward force created by the fusion reaction, its own gravity had crushed it to the size of a planet. The question then became: what stopped gravity from crushing it even further? Answering that question would set scientists on an unstoppable march to black holes.

The extraordinary thing about the 1920s was that, in no small part due to Einstein, physicists now had a comprehensive set of tools with which to probe the behaviour of matter. At the scale of the very big, they used the laws of General Relativity. At the scale of the very small, they had quantum physics. In 1928, a nineteen-year-old Indian student helped himself to this shiny new toolkit to come up with the answer as to why a white dwarf that had run out fuel was nevertheless able to resist gravity.

Subrahmanyan Chandrasekhar was on a slow boat to Southampton, en route to Cambridge to start a PhD. Eager to impress his teachers, he read Eddington's textbook and was enthralled by Sirius B and its incredible density. Deprived of fusion energy, nevertheless something about this new state of matter allowed it to resist the crush of gravity. The answer appeared to lie in another paper he read on the boat by Cambridge physicist Ralph Fowler. It described how the new laws of quantum mechanics applied to extremely dense matter. It turned out that Sirius B could still defy gravity's crush thanks to the wave-like properties of close-packed electrons – so-called "electron degeneracy pressure".

During Chandrasekhar's eighteen days at sea, the whizzkid deftly juggled classical physics with the twentieth-century game-changers of relativity and quantum mechanics, and he arrived at a remarkably simple conclusion. The wave pressure of crowded electrons could resist gravity – up to a point. The limit was a mass equivalent to 1.4 suns. Fifty-three years later, Chandrasekhar would win the Nobel Prize for divining what is now known as the Chandrasekhar limit.

Chandrasekhar had shown the maximum size for a white dwarf. What would become of stars more massive than that? The universe was full of giants, such as Betelgeuse, the star that marks Orion's shoulder. Its mass is 10–20 times that of our sun. What would be the fate of a star as big as Betelgeuse?

In 1933, California Institute of Technology physicist Fritz Zwicky guessed.[24] It would be a star composed entirely of particles known as "neutrons". It was a breathtaking leap on his part. The neutron had only been discovered a year before. They were believed to form when a positively charged proton fused with a negatively charged electron. Zwicky imagined himself inside a giant star that had run out of fuel and was being crushed by its own gravity. Inside this gargantuan compactor, electrons that had once luxuriated in vast amounts of space would become sardined against each other. In stars whose final mass was greater than 1.4 suns, the compactor would continue until electrons collapsed into the core and fused with protons, producing a star composed entirely of neutrons. It would be at least as dense as the fantastically dense nucleus of an atom, 230 trillion tonnes per cubic metre.

Zwicky's leap suggested an answer to two other cosmic mysteries: the origin of rare stars known as supernovae that suddenly increased their brightness a billion-fold, and cosmic rays, mysterious high-energy particles from space. Zwicky proposed that prior to its final collapse into a neutron star, the massive star would explode as a supernova, and that the ejecta from such explosions were the source of "cosmic rays" (now known not to be rays, but mostly hydrogen atoms stripped of their electrons, shooting through the universe at close to the speed of light).

Decades later, Zwicky's cosmic leap would turn out to be dead right. But at the time, few of his colleagues paid much attention to his theory, in part because Zwicky's fiery temperament did not engender confidence. "Many of Fritz Zwicky's colleagues regarded him as an irritating buffoon," writes Thorne.[25] More importantly, there was no maths to back up his notion that a star could exist if composed entirely of neutrons.

Those missing calculations came from an imperilled young Moscow-based mathematician, Lev Landau. His motivation was not entirely academic – he was desperate to publish a high-profile paper in the hope it would protect him from the Stalinist purges of 1937. He succeeded in publishing a paper in *Nature*, but not in escaping a year in prison, which he barely survived. (He would nevertheless go on to win a Nobel Prize for cracking the mystery of superfluidity.)[26]

Landau's paper, "The Origin of Stellar Energy", was read with great interest by Robert Oppenheimer, a theoretical physicist at the University of California, Berkeley. Not knowing that the author was languishing in a notorious Moscow prison, Oppenheimer rejected some of Landau's conclusions but was persuaded by his calculations showing that neutron stars could exist.

Using Landau's maths, Oppenheimer modelled the fate of a collapsing star. If it was more massive than 1.4 suns, gravity would crush it into a neutron star. But was there a maximum size for a neutron star? Oppenheimer's student George Volkoff calculated that neutron pressure could resist gravity for a star up to six times the size of the sun.[27] In other words, the maximum size of a neutron star was six solar masses. (Today, physicists believe it is closer to two.[28]) What, then, was the fate of even larger stars? Modelling what happens in an imploding star is diabolically complex. Oppenheimer and another student, Hartland Snyder, simplified that complexity to a "stripped-down" model: a perfectly spherical, non-rotating, imploding star.[29] A real imploding star would *not* be a perfect sphere and would in fact rotate.

They published their model on 1 September 1939 in *Physical Review*.[30] The massive star would continue to crunch until it reached a singularity – an infinitesimally small point that we now know as a black hole. Inside it, all matter would be shredded into something mysterious. No theory of physics exists to describe it. Furthermore, an outside observer would never see the star proceed to the final crunch. That's because its immense gravity, concentrated over a tiny volume, would twist spacetime – and, according to the laws of General Relativity, cause time to slow down and eventually stop.[31]

Few of Oppenheimer's colleagues believed this "stripped-down" model had any relation to reality. More likely, giant stars, which exploded as supernovae prior to their final implosion, would lose so much mass that they would never end up as large as the star in Oppenheimer's model. And even if they did, the laws of General Relativity would probably fail at such extremes.

Oppenheimer and Snyder's paper marked the end of an eleven-year period when physicists had combined the laws of the very big and the very small to predict the fate of large stars. That enquiry came to a halt the day their paper appeared in *Physical Review*. Not because they had solved the mystery of black holes, but because physicists suddenly found themselves in a black hole much closer to home. On 1 September 1939, Hitler invaded Poland, an event that would come to mark the start of the Second World War.

Father of the atomic bomb

For physicists, like everyone else in 1939, the focus was now very firmly back on earth. Einstein had flagged an existential threat a month before, in a letter to US president Franklin D. Roosevelt: Germany could well be constructing an atomic bomb.[32]

In January of 1939, German physicist Otto Hahn had announced something many considered impossible: the nucleus of an atom could be split. Previously, indivisibility was considered the very definition of the atom. My friend Bella Hirshorn recalls learning it as a high school student in Wilno, Poland, in 1937. The word "atom" derived from the Greek *atomos*, which means "indivisible". The Nobel Prize–winning Danish physicist Niels Bohr broke the astonishing news at a Washington conference. He also shared the fact that splitting the uranium-235 atom released huge amounts of energy. The danger was clear to Leo Szilard, a Hungarian-born Jew who had trained in Germany and was now a refugee in the United States. In 1933, he had predicted the possibility of a nuclear chain reaction and its destructive power. Now, that feat had been achieved by a country with a leader whose stated ambition was world domination.

It was Szilard and Eugene Wigner, another Hungarian-born Jewish physicist (and future Nobel Prize winner), who persuaded Einstein to write the letter. Ultimately, that letter triggered Roosevelt to instigate a top-secret mission to develop an American atomic bomb. In June 1942, Oppenheimer (also of Jewish descent) left black holes behind for the Los Alamos lab in a remote part of New Mexico, where he became scientific leader of the Manhattan Project.

There was every reason to believe the Germans were way ahead in the development of an atomic bomb. With the likes of Max Planck, Erwin Schrödinger and of course Einstein and his equation $E=MC^2$, which showed how matter might be converted to energy, Germany had been the birthplace of atomic physics. And although Jewish and liberal-minded physicists had fled, pioneers such as Otto Hahn, Werner Heisenberg and Hans Geiger remained and were put to the task.[33]

Even with its headstart, however, the German nuclear weapons project fizzled out. Many of its key scientists were more motivated to develop atomic science as a future energy source than to build a bomb for the Führer. Heisenberg may even have sabotaged the German nuclear effort. More generally, it was unwise to overpromise on a project to a hot-tempered leader with a general disdain for science – especially the so-called "Jewish science" of nuclear physics. According to US historian McGeorge Bundy, "From 1939 onward, the dominant motive of most German physicists had been to assure the protection of both their science and their scientists from Hitler's war … There is a certain decency in what they did and did not do."[34]

On the other hand, secret tape recordings from Farm Hall, the open prison that the German nuclear bomb team were taken to, show that when they heard about Nagasaki and Hiroshima, they were astonished, "as if they genuinely believed it was impossible to make a practical bomb", notes theoretical physicist Bernard Schutz.[35]

Another American physicist recruited to the war effort was John Wheeler. As a junior scientist, he had worked with Niels Bohr at

Princeton in 1939 to figure out how to start an atomic chain reaction by firing neutrons, like cannonballs, at uranium-235, to split it open and release more neutrons. Wheeler applied that understanding to the development of an atomic bomb – with special poignancy. His brother, a PhD in American history, was fighting in northern Italy. His last two words to Wheeler came by postcard: "Hurry up."[36]

Wheeler became the lead scientist for the design and construction of the world's first production-scale nuclear reactors for plutonium-239, the major fuel for nuclear weapons, in Hanford, Washington.[37] (Unlike uranium-235, which is mined, plutonium-239 must be manufactured.)

As Hiroshima and Nagasaki attest, American physicists succeeded. Some were mortified at the apocalypse they'd unleashed and feared for the future of humankind. "Physicists have known sin," declared Oppenheimer.[38] "The unleashed power of the atom has changed everything save our modes of thinking, and thus we drift toward unparalleled catastrophe," warned Einstein.[39]

Einstein and Szilard, the authors of the letter that triggered the Manhattan Project, devoted the rest of their lives to averting that catastrophe through activism against nuclear weapons proliferation.[40] Oppenheimer atoned by pushing for international arms control as chairman of a committee set up by the Atomic Energy Commission and returned to Princeton to head its Institute for Advanced Study.[41]

But while many atom bomb scientists returned to civilian science, John Wheeler was drawn back into the Cold War efforts to build a "super bomb". America had panicked when the USSR tested its first atomic fission bomb in August 1949. Edward Teller, a Hungarian colleague of Leo Szilard, argued that the US must secure its dominance by building a super bomb. Instead of fission, it would employ star power – fusion.

Oppenheimer initially opposed the idea, both on moral grounds and because it seemed technically unfeasible, arguing that "the extreme dangers to mankind in this proposal wholly outweigh any military advantage that could come from this proposal".[42]

Wheeler took a different view and agreed to lead the fusion program at Princeton. Superbombs became a reality. On 1 March 1954, the US exploded a hydrogen bomb with the destructive power of 1300 Hiroshimas. In 1961, the Soviets detonated the equivalent of 5000 Hiroshimas.

Superbombs were on their way to wiping out the entire human race. As anti-bomb campaigner Howard Morland put it:

> Any cognitive species which begins to comprehend the relationship of matter and energy will shortly discover that sunshine, the energy source of life, is the output of a thermonuclear furnace. With that knowledge, plus a few engineering innovations, the entire species will collectively confront Hamlet's dilemma: "To be, or not to be."[43]

Total destruction of the human race does not make for a useful tactical weapon. The H-bomb project was ultimately wound back.

From bomb maker to black-hole hunter

Wheeler was finally able to return to academic research in March 1956. He picked up where he and Oppenheimer had left off seventeen years before. What was the fate of large stars? In his view, "Of all the implications of General Relativity for the structure and evolution of the universe, this question of the fate of great masses of matter is one of the most challenging."[44]

Wheeler did not come to accept black holes lightly. He had been a strident critic of Oppenheimer's stripped-down 1939 model. But in 1939, the mathematical tools to compute all the messy details of an imploding star – the pressures, nuclear reactions, shock waves, radiation and mass ejection – didn't exist. In the intervening decades, those tools were provided by the nuclear weapons design effort. As Thorne put it, they "are all central features of a hydrogen bomb".[45]

Wheeler gave the task of simulating the fate of massive stars to Masami Wakano, a postdoctoral candidate from Japan. Wakano had access to MANIAC, one of the world's first digital computers. Filling an entire room at the Princeton Institute for Advanced Study, it had been built to crack the codes for hydrogen bombs. In a matter of hours, it did what Oppenheimer's students had taken days to do. Wakano's calculations, presented at a conference in Belgium in 1958, vindicated and refined Oppenheimer's findings of twenty years before. Stars smaller than 1.4 solar masses ended up as white dwarves. Stars whose remnants after exploding as supernova were between 1.4 and two solar masses ended up as neutron stars. But if their remnants were larger than two solar masses, there was only one fate: a black hole.[46]

Wheeler: black holes to gravitational waves

And so, decades before there was a shred of astronomical evidence, the maths told Wheeler that black holes must be real.

It reset the trajectory of Wheeler's career. The study of general relativity and gravitation, he later wrote, was "the second half of my life". Wheeler was also a famously dedicated teacher, remembered by one student, Kip Thorne, as "the most influential mentor of young scientists whom I have known".[47]

After their first meeting, Thorne was hooked on black holes and joined the ranks of a whopping forty-six PhD students and other scientists to have been mentored by Wheeler. They included PhD student Richard Feynman, who would push the frontiers of the quantum world and earn a Nobel Prize. PhD student Hugh Everett would fearlessly develop the theory of multiple worlds. And the equally fearless Joseph Weber, who spent a year in Wheeler's lab, would be the first to build a device to detect gravitational waves. As a PhD student, Thorne would build the theoretical edifice for how to decode such waves and ultimately win the 2017 Nobel Prize for their detection.

So, two decades after Einstein tried to do away with them, how did gravitational waves come back into the picture? Wheeler came to believe that the best way to detect those black holes would be using gravitational waves. The deductive pathway was simple. Unlike our sun, most stars have a dance partner. They waltz around each other in what is known as a "binary system". In some cases, one or both partners end their starry lives as a black hole. The waltz continues – but more intensely. Their immense gravity concentrated to a single point churns up spacetime. Unlike normal binaries, their orbits shrink; they start spiralling, and eventually they crash into each other. During this last waltz, the binary would radiate titanic gravitational waves. And Wheeler, unlike Einstein, believed they would be detectable.[48]

That belief was forged at a conference on gravity held in North Carolina in 1957.[49] It was held to inaugurate the new Institute of Field Physics headed by Bryce DeWitt. Delightfully, he'd landed the position thanks to stock market millionaire Roger Babson, who attributed his fortune to the laws of gravity: what goes up must come down. The star performance came from Richard Feynman. A former PhD student of Wheeler's but an outsider to general relativity, he nevertheless presented a thought experiment that demonstrated that gravitational waves "are in every respect real."[50]

Building the theoretical edifice

While Wheeler inspired the quest to find gravitational waves, Thorne became one of its most dedicated foot soldiers, setting up a lab at Caltech in 1966 to further the cause. Joseph Weber had constructed aluminium bars to resonate in sympathy with a passing gravitational wave. Since Weber had laid the experimental foundations, Thorne would lay the theoretical ones. There was no time to lose; the aluminium bars were already ringing. Thorne's goal was to decode the information in gravitational waves as surely as your computer modem decodes radio waves into text, images and sound. In this case, however, decoding gravitational waves would reveal the choreography of waltzing black holes: how fast they circled each other, how massive they were, and how far they were from earth.

On the face of it, it seems a straightforward problem. Say two black holes are whizzing around each other at 100 times per second. Like propeller blades in water, their spinning churns up spacetime to create gravitational waves with a frequency of 100 cycles per second. So to a first approximation, the frequency of gravitational waves reflects the speed at which two black holes are spinning around each other.

However, there's a lot more to it than that. For one thing, their orbits change as their masses spiral inwards. The calculations required to model this scenario are dauntingly complex. In the 1960s, supercomputers attempting such simulations crashed. "Even today, it's still very computationally expensive to simulate a few tens of orbits, plus the collision, of two objects," explains Rory Smith, a onetime PhD student of Kip Thorne, based at Monash University.[51]

In the 1960s and '70s, Thorne also collaborated with theoretical physicist Yakov Borisovich Zel'dovich, whose group at the Keldysh Institute of Applied Mathematics in Moscow from the mid-1960s to early 1970s, he described, as "superior to any other theoretical astrophysics team anywhere in the world".[52] Zel'dovich was Wheeler's mirror image: he headed the Russian H-bomb project and then moved into black hole research. His colleague Vladimir Braginsky, inspired by Weber, was also building resonant bars.

Thorne and Zel'dovich were erecting a dazzling set of theoretical foundations for the behaviour of waltzing black holes – all without a shred of evidence that black holes actually existed. Their confidence was based on Einstein's General Relativity, the predictions of which had so far proven bulletproof. Moreover, their calculations were continually being crosschecked by different mathematical tests. The process was self-correcting in the same way that you know you have the right solution for a crossword puzzle when each word fits the requirements of multiple columns and rows.

Still, it is possible to get a crossword puzzle wrong. Some actual astronomical evidence for the existence of black holes or neutron stars would have been a welcome sign from the heavens.

A few years after Thorne got started, three such promising signs appeared.

Heavenly signs

In the early 1960s, research rockets were blasted into space to study space X-rays. Their onboard Geiger counters also happened to pick up the atmospheric fallout of the 1961 Russian hydrogen superbomb test over the Arctic, the most powerful bomb ever exploded. But that signal was nothing compared to the signal coming from the constellation Cygnus X-1 in the Milky Way. What could be responsible? A black hole, thought Thorne. It was just the signature that Zel'dovich had predicted. As gases were sucked in at close to light speed and heated to millions of degrees, they would radiate high-energy X-rays. Those radiations from Cygnus X-1 were the first hint that black holes were real.

The next appeared in 1967, when Jocelyn Bell, a PhD student at Cambridge, was listening to radio waves. These metre-long, low-energy waves are at the other end of the spectrum to billionth-of-a-metre X-rays. Bell was listening to the skies over Cambridge using an array of radio antennae. To her astonishment, she heard a regular pulse coming from the constellation Vulpecula, or the fox.

Little green men, was her initial thought. And so LGM-1 was the name she gave to the signal which beeped every 1.3 seconds, like a metronome. Within a couple of months, she found three more of them. Rather than little green men, there was an alternative explanation. It was no less thrilling. These were the broadcasts of a massive, compact spinning star whose day could be measured in seconds. Bell and her PhD advisor, Antony Hewish, published a paper in *Nature* in 1968, proposing: "The radiation may be associated with oscillations of white dwarf or neutron stars."[53]

It had taken thirty-five years, but Fritz Zwicky's intuition had been vindicated. Stars more massive than 1.4 suns would end their days as neutron stars. The pulsating signal turned out to be that of a fantastically dense neutron star. A teaspoon of its mass would weigh 10 million tonnes![54] Its magnetic field was hundreds of millions of times stronger than that of Earth. And as it spun, it emitted a beam of radio waves, sweeping across space like a lighthouse beacon. Bell's Cambridge radio antennae were fortunate enough to lie in its path. The unfailingly accurate pulses from these pirouetting neutron stars were dubbed "pulsars".[55]

The third big sign from the heavens came in 1974. Neutron stars not only pirouetted; they formed waltzing pairs. And for the real crowd-pleaser: as they waltzed, they radiated gravitational waves.

Russell Hulse, a PhD student at the University of Massachusetts Amherst, made the discovery using the Arecibo radio telescope in Puerto Rico. Built into a natural sinkhole, it was then the world's largest 'dish' and allowed him to find pirouetting pulsars in the Milky Way at the bonanza rate of one every ten days. Then he found a very unusual pulsar. Its metronome was *irregular*. The reason soon became clear to Hulse and his advisor, Joseph Taylor. The glitchy pulsar was waltzing around another neutron star. As it waltzed away from Earth, its frequency slowed for the same reason that an ambulance siren's pitch lowers as it speeds away from you – the Doppler effect. Likewise, as the pulsar waltzed back towards Earth, its frequency increased like that of an approaching siren.

This binary pulsar had one more fantastic secret to reveal. The radius of its waltz appeared to be shrinking at about 3.5 metres per year. That meant the two neutron stars were spiralling inwards towards each other. This was just what Wheeler had imagined black holes might do. Moreover, they were getting closer to each other at just the rate predicted if they were emitting gravitational waves.[56]

There was no other explanation, said Thorne. It was not a direct detection of gravitational waves, but it was the first piece of evidence for their existence.[57] Hulse and Taylor would win the 1993 Nobel Prize for their discovery.

Doing science backwards

By 1976, Thorne was so confident gravitational waves could be detected from the collisions of neutron stars or black holes that he persuaded Caltech to build a gravitational wave detector. He hoped it would complement the efforts of "more than a hundred talented experimenters" who were patiently listening to resonant bars in universities across the world, as well as the Bell and IBM labs.[58]

Ronald Drever, the Scottish physicist appointed to build the detector, had no wish to build another resonant bar. He had listened to them at the University of Glasgow for five years without hearing anything but noise. He was keen to build an interferometer, an instrument 10,000 times more sensitive than a resonant bar. He had begun developing a prototype. So had physicist Rainer Weiss at the Massachusetts Institute of Technology (MIT).

But there was a hitch. While resonant bars could be built and operated by a lone scientist, a working interferometer was an industrial proposition. It would require the engineering of perfect lasers, mirrors and vacuum tunnels, and the development of new electronics and software. By 1984, the National Science Foundation was willing to fund the ambitious project. On one condition: the merging of Drever's Caltech and Weiss's MIT teams, each comprising about eight people. It was, as Thorne recalls, a shotgun marriage. "Drever (resisting like mad) and Weiss (willingly accepting the inevitable) said their vows and I became the marriage counsellor."[59] And thus, the Laser Interferometer Gravitational Wave Observatory, or LIGO, was born. The marriage was not a success. In 1986, CEO Robbie Vogt was appointed to help transform the fiercely individualistic culture to one that would foster a collaborative spirit.

Meanwhile, theoretical physicist Bernard Schutz at the Cardiff University, a former student of Thorne, provided a tantalising incentive for the building of the instrument. As he outlined in a 1986 paper in *Nature*, gravitational waves offered a new way to measure the rate at which the universe is expanding.[60]

Edwin Hubble first realised that our universe is expanding in 1929, when he focused his telescope on twenty-four galaxies. He found that the further away they were, the faster they travelled. This expansion rate, known as the Hubble constant, is the key to calculating the age of our universe. Winding back the timer should take us to the beginning, when all the universe was, according to Stephen Hawking, concentrated at a singularity.[61]

So far, estimates of the universe's expansion rate differ. One reason is that measuring the distance to other galaxies relies on using "standard candles", stars of known brightness. The dimmer the standard candle, the further away its galaxy must be. But standard candles are not as "standard" as astronomers had hoped, so calculations based on them have produced inconsistent results.

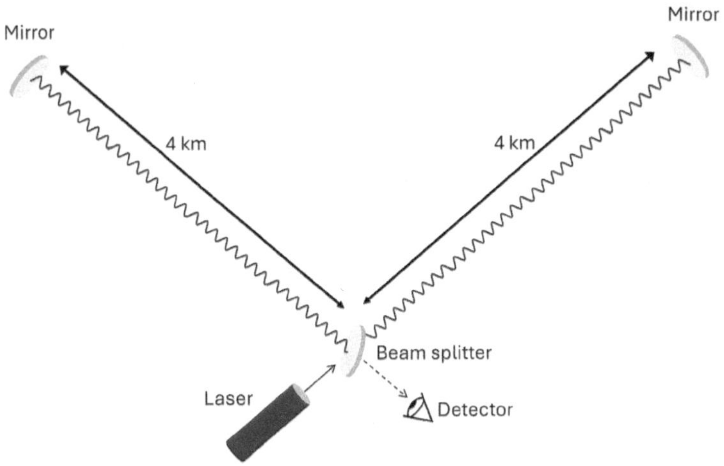

Figure 14. To catch a gravitational wave …

Just as a passing ocean wave creates peak and troughs in the water, a gravitational wave distorts spacetime, stretching it in one dimension while squeezing it in the other. The Laser Interferometer Gravitational Wave Observatory (LIGO) is designed to measure these dimensional changes, which are unfathomably small: around one ten thousandth the diameter of a proton (10^{-19} metre). Extreme sensitivity means the interferometers are susceptible to noise, such as the rumble of a passing truck ten kilometres away. That's why there are twin detectors on opposite sides of the USA, in Washington State (Hanford) and Louisiana (Livingston). A signal is only taken seriously if both detectors pick it up within milliseconds of each other (the time it takes for a wave to travel the 3000 kilometres between them).

How the interferometer operates: A laser fires a light beam, and the beam splitter sends one stream off at right angles while another passes straight through. Mirrors reflect the beams back to the splitter. The beams are exactly half a wavelength out of phase, so normally they cancel each other and there is no signal to detect. A passing gravitational wave will stretch one arm while squeezing the other, putting the laser beams out of phase. Eureka! A signal will be detected.

Schutz showed that gravitational waves would deliver an independent way to measure cosmic distances. Just as the size of a sound wave is related to the distance it has travelled, the size of the gravitational waves would represent how far they had travelled from the colliding black holes or neutron stars that created them. In other words, gravitational waves would serve not as "standard candles" but as "standard sirens". That promise rallied the astronomical community behind the LIGO project, and funding flowed.

By 1990, the plans to construct LIGO were approved and, two years later, sites for the twin instruments, which would operate in unison, were chosen – 3000 kilometres apart, on diagonally opposite sides of the USA, in Hanford, Washington and Livingston, Louisiana. From 1994, the project was led by physicist Barry Barish, a veteran of big science projects who would eventually share the 2017 Nobel Prize with Thorne and Weiss.[62]

With Barish at the helm, the construction of LIGO surged forward and Thorne was seized by an awkward fear. The machine might detect gravitational wave signals before his cabal of theoretical physicists had figured out how to decode them.[63] When Thorne had begun his quest thirty years before, he had no idea how monumental the challenge would be.

Traditionally, physicists start with a real phenomenon, such as a tidal wave, and measure physical characteristics, such as height and frequency. From these they can infer information about the quake that caused them. But this new project was doing science backwards. You "believe" there are colliding black holes out there. You want to detect the reverberations coming from the tsunami they create. And you want to capture all the information carried by those waves, such as how distant the collision was and the size of the black holes. If this were a study of ocean tsunamis, you could test your models using a historical data set.

In the absence of real data, however, Thorne had to rely on simulations. The problem was, the simulations of the 1960s and '70s applied to a "stripped-down" model: a head-on collision of two black holes.[64] By the 1990s, it was clear that something far more complex would take place between a pair of black holes in space. Take a look at this description from a group of theoretical physicists at the Simulating eXtreme Spacetimes project, or SXS:

> As each [black hole] spins, it drags the spacetime around it into a tornado-like whirl. As the pair orbit around each other, the holes' orbital angular momentum creates a third, larger whirl. What happens when two tornadoes, embedded in a third larger tornado, come crashing together and the tornadoes are made from whirling space, not whirling air?[65]

In 1995, physicists did not have the tools to model tornadoes in spacetime. To develop them, physics gave rise to a new priesthood – "numerical relativists" – who understood the esoteric maths of general relativity and could program supercomputers to deploy it.

Just as Thorne had feared, the experimentalists raced way ahead of the theorists. By 2002, LIGO was up and running, searching for gravitational waves. But the numerical relativists were still crashing their supercomputers when they tried to simulate colliding black holes.

Finally, in June 2005, the SXS achieved a workable simulation thanks to a breakthrough by one Frans Pretorius. A year later, two independent groups improved on those simulations. One led by Joan Centrella at Goddard Space Flight Center and the other by Manuela Campanelli at Rochester Institute of Technology – "two women in what was largely a male priesthood", notes Bernard Schutz.[66] They used and improved an open-source computer code that had been developed at the Albert Einstein Institute in Potsdam, Germany, where Schutz was the founding director. "So they beat us to the result using basically our own code."

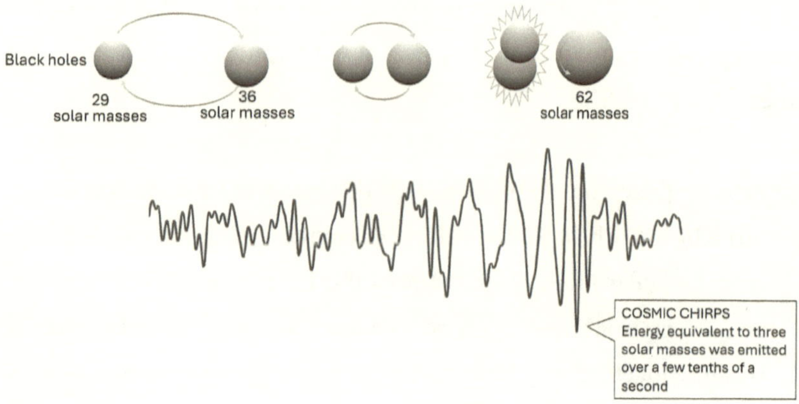

Figure 15. The last song of waltzing black holes. Circling each other, black holes emit gravitational waves. As they grow closer, the frequency and height of the waves increase until in a final crescendo they collapse into each other. All the information about the masses of the waltzing black holes, and that of the final collapsed one, is contained in the shape of the waveforms. But it had to be deciphered by the simulations of theorists.

Ultimately, LIGO had access to around a thousand simulations, contributed by different groups. These simulations described a catalogue of black holes, each with different masses, spin rates, spin angles and distances from Earth. LIGO used this catalogue the way a birdwatcher might use an app to identify an unseen bird from its call. Without the app's catalogue of calls, the listener could still enjoy the beautiful song, but she would remain clueless as to the singer.

The rest is history

In 2015, the catalogue was finally field-tested.

LIGO's first scan of the universe, from 2002 to 2010, had failed to snare a gravitational wave. On 14 September 2015, the updated LIGO was going through a test run. With a ten-fold increase in sensitivity, it could now scan a volume of space 1000 times greater than its predecessor.

Two days after turning the machine on, the two detectors, on opposite sides of the American continent, almost simultaneously registered a wave. As rare birds go, this one was a beauty. And thanks to the catalogue, its call could be identified.

Named GW 150914 for having been discovered on the fourteenth day of the ninth month of the year 2015, it had been created by the collision of two black holes, one weighing twenty-nine solar masses, the other thirty-six solar masses. Their merger created a black hole weighing sixty-two solar masses (three solar masses were converted to energy and radiated by the gravitational wave). When the wave signature was played through a speaker that translated the gravitational wave into a soundwave, it produced a high-pitched chirp.

What was particularly uncanny was how the wave fit the simulation. Perfectly. So perfectly, in fact, that the first person to lay eyes on it didn't believe it. Around noon, Marco Drago, a 33-year-old postdoctoral researcher at the Albert Einstein Institute in Germany, received an automated email alert. The software sifting through the LIGO data had just detected an anomaly. Clicking on the hyperlink, he saw a shapely squiggle poking out above the noise, growing bigger and bigger, faster and faster.[67]

Drago, who moonlights as a sci-fi novelist, had many reasons to be suspicious. For one thing, the revamped machine was still having its tyres kicked during "engineering runs". For another, a small, secretive group of LIGO members had a licence to trick the data analysts with so-called "blind injections". This might involve shaking the mirrors in the machine to simulate a wave, or monkeying with the data.

However, when Drago finally got through to the LIGO team at Livingston at 5.30 am, the operator knew nothing about a "blind injection". The LIGO chiefs went on high alert. "Í looked at this thing and thought, my god, this looks like it," said Thorne.[68] Either the signal was real or there was an anonymous prankster at large. Both possibilities were concerning. If it was real, the LIGO team did not want that to leak just yet. They needed to check all the measurements and do further statistical analyses. They would not go public till their analysis had passed the high statistical bar of five sigma, a one in a 100,000 chance that it could be due to noise. And their work had to be peer-reviewed.

There was a reason for the extreme caution. Just nine months before, a group of radio astronomers had to retract a claim that their telescopes had detected the signature of gravitational waves in the skies over Antarctica. Humiliatingly, the swirling signatures they had identified as gravitational waves from the Big Bang turned out to be nothing more than swirls in a cloud of Milky Way dust.[69]

But the LIGO results were real.

The press conference to announce them, held the morning of 11 February 2016 at the Washington Press Club, had the gravitas of a once-in-a-century event. With barely suppressed emotion, David Reitze, the LIGO executive director at Caltech, announced: "Ladies and gentlemen, we have detected gravitational waves. We did it." It was a proof 100 years in the making, built on the backs of generations of physicists, helped along by bomb-makers, astronomers, bold experimenters, numerical relativists and teams of gifted engineers.

This was not just the first detection of gravitational waves; it was also the first direct detection of the existence of black holes. Like General Relativity, black holes had been deduced from the workings of a human mind, a century before the experimental tools existed to detect them. Stephen Hawking's words capture the sense of awe:

> I do not know any other example in science where such a great extrapolation was successfully made solely on the basis of thought. It shows the remarkable power and depth of Einstein's theory.[70]

Next

To date, LIGO has picked up some ninety gravitational waves, most of them from colliding black holes, but some from colliding neutron stars.[71] That *so many* black holes exist has been astonishing. "The universe is packed full of black holes!" says theoretical astrophysicist Susan Scott, who explores black holes from the Australian National University. Astronomy has entered a new era, routinely detecting phenomena that were once considered as fanciful as the flapping of an angel's wings. In 2017, a detector in Pisa, Italy, joined in the quest and next-generation detectors are being planned around the world.

The ability to detect gravitational waves from colliding neutron stars has also provided other trophies. For one thing, these collisions, unlike the dark collisions of black holes, are detectable by telescopes. Peering at the explosion caused by colliding neutron stars on 17 August 2017, astronomers were able to crack another mystery of the universe.[72]We have long known that elements from helium to iron were created by the fusion of lighter atoms in the core of large stars. And the furnace for heavier elements had been traced to supernovae. But the source of even heavier metals, such as gold and platinum, remained a mystery. It was in the fiery explosion of colliding neutron stars – the so-called "kilonova" – that astronomers first saw the chemical signatures of gold and platinum. The quantity of gold created in the collision was estimated to be 40–100 times the mass of Earth.[73]

The same event also allowed two more tests. Einstein's equations had predicted that gravitational waves would travel at the speed of light. Given the light waves were detected just 1.7 seconds after the arrival of the gravitational waves, over a journey of 1.3 billion years, that was considered close enough![74] The second test was a new estimate of the Hubble constant, just as Schutz predicted.[75]

But the most fundamental test was that Einstein's theory of General Relativity held up under one of the most extreme conditions in the universe: the collision of two black holes. As a result, "we can rule out other theories of gravity", says Scott.[76]

Despite all these triumphs, today's physicists are hardly strutting around, slapping each other on the back for proving Einstein's theory. It is the exact opposite. They seem to be growing bored with the sport of proving General Relativity. Their great ambition now is to be part of its undoing – to take their place in the ongoing pageant of physics, where paradoxes abound and may reveal the next secret of the universe.

Take the singularity inside a black hole. As Wheeler realised seventy years ago, the pillars of twentieth-century physics, General Relativity and quantum mechanics, will not stand inside a black hole. Instead, a new theory, dubbed "quantum gravity", will describe the nature of spacetime inside a singularity. Will it be scrubbed into a quantum foam, as Wheeler imagined? Quantum gravity also predicts gravity particles, or "gravitons", akin to the photons of electromagnetic waves. The next generation of LIGO detectors may be able to detect them in the noise, suggests Scott.

"John Wheeler tried to leapfrog his way into the twenty-first century to catch a glimpse of what the laws of physics might be," reflected Thorne. This generation of physicists should now be able to reveal them.

Watch this space.

Epilogue

The idea for this book was hatched in 2019, during the first Trump presidency, as a response to the attack on truth. Now, as I write this epilogue in early May 2025, the first hundred days of Trump's second presidency have passed and it is not just truth but science that is under attack.

The scope of the carnage is hard to comprehend. Nothing seems to have been spared. Not NASA, not NOAA (ocean and atmospheric research), not the Environment Protection Authority, not the Department of Agriculture, which fired bird-flu experts in the middle of an outbreak.[1] USAID, which delivered the US President's Emergency Plan for AIDS Relief, a program that provided HIV testing and drugs, is now gone. The government contribution to a public–private program to fund global vaccines, GAVI, is also under threat.[2] These inhuman acts will cost millions of lives.

Elite research universities, including Harvard, Columbia, Princeton and Cornell, are prime targets. Nor have the crown jewels of American science – the National Institutes of Health and the National Science Foundation – been spared. The $47 billion budget of the NIH, the largest funder of biomedical research in the world, faces a 40 percent cut. The White House's rationale is that the NIH "has grown too big and unfocused" and "has broken the trust of the American people with wasteful spending, misleading information, risky research, and the promotion of dangerous ideologies that undermine public health".[3]

That "wasteful spending", together with about half as much from the pharmaceutical industry, contributed to 99 percent of the drugs approved in the United States from 2010 to 2019, including cancer immunotherapy, heart attack and stroke medications.[4] The NIH led the fight against Covid-19, developing rapid test kits and carrying out the first clinical trial for a vaccine as well as trials for "off-shelf" drugs repurposed for Covid.[5] Ten leading researchers were summarily sacked, including Richard Youle, who won the 2021 Breakthrough Prize for his work on Parkinson's disease.[6]

As for the National Science Foundation, "Every current technology has the imprint of NSF investments over the past decades. NSF is truly the envy of the world," wrote Sethuraman Panchanathan in his resignation letter on 24 April, after being told to slash the NSF's $9-billion funding and staff by half.[7]

"It's hard to overstate how serious this is … It's hard to believe any administration would do this," noted Alan Bernstein, director of global public health at Oxford University.[8] "These are the building blocks of America's extraordinary strength, created over the last 100 years. They are now being dismantled in just 100 days," wrote Fareed Zakaria in *The Washington Post*.[9]

As far as projects, any research that mentions Diversity, Equity and Inclusion (DEI), climate change or vaccine hesitancy is a clear target, but some 1000 research grants have also been cancelled across cancer, Parkinson's disease, Alzheimer's disease, tuberculosis, HIV prevention and long Covid.[10]

Pandemic preparedness has been dismantled. Projects to develop a generic vaccine against the next pandemic coronavirus, develop better drugs against Covid and other threats such as Ebola and Zika, have been cancelled. It is a "pointless, ill-advised move that will hurt US science and pandemic readiness," says Charles Rice, a Nobel Prize–winning virologist at the Rockefeller University who co-leads some of this research.[11]

Experts have been summarily fired and replaced by sycophants. The Department of Health and Human Services, which oversees America's health infrastructure as well as the Food and Drug Administration, the Centers for Disease Control and the NIH, is now led by America's most prominent anti-vaxxer, Robert F. Kennedy, Junior.[12] One of his first acts was to force out Peter Marks, the FDA's chief vaccine regulator, who approved the monoclonal antibodies that saved Trump's life.[13] RFK also sacked 10,000 employees whose work covered everything from preventing childhood lead poisoning to HIV prevention and food safety. "Devastating, thoughtless, chaotic – I don't think they fully understand what they've done," said David Kessler, a former FDA chief. "Bad stuff is going to happen on their watch. They're not going to be equipped."[14]

Kennedy visited Samoa in 2019, when only a third of the population were vaccinated against measles. Alec Ekeroma, Samoa's director general of health, described Kennedy's visit as a "significant disinformation campaign" to stoke distrust in vaccines. Four months later, a measles outbreak took eighty-three lives, most of them babies and young children.[15]

Now Kennedy presides over the worst US measles epidemic since the country declared the disease eliminated in 2000. As of 2 May 2025, across twenty-seven states, there are 935 cases, of which 13 percent have been hospitalised. Two children and one adult have died, all of them unvaccinated, as are 96 percent of the cases.[16] Kennedy is a champion of the theory that the MMR vaccine causes autism, a theory that has been thoroughly tested and disproved.[17] So while cutting "wasteful spending" at the NIH, Kennedy has commissioned a new study to be led by a longtime opponent of vaccines with no scientific credentials who has practised medicine without a licence.[18]

"We will learn absolutely nothing with another study," says Helen Tager-Flusberg, director of the Center for Autism Research Excellence at Boston University. "We've ruled it out."[19]

Kennedy promotes unconventional treatments for measles such as vitamin A, steroids and antibiotics and says measles vaccination "is a personal choice".[20] Besides Kennedy's continuing efforts to sow distrust, vaccination clinics across the country are being forced to shut down because of the $11-billion cuts to federal funding.[21]

The justice system, the media and other institutions of government are also victims of the post-truth era. As I write, two US judges have been arrested for opposing summary deportations.[22] Trump has signed an executive order to stop federal funding for public broadcasters NPR and PBS, brought lawsuits against the media and denied the Associated Press access to the Oval Office.[23] The National Security Agency lost its director and deputy director after far-right influencer Laura Loomer denounced them to Trump as insufficiently loyal.[24]

In his book *The Constitution of Knowledge: A Defence of Truth*, Jonathan Rauch notes that all these organisations – academia, law, journalism and government – need to operate based on a shared reality.[25] To do so, they all employ the scientific method, a method that tests every theory and relies on facts. But Trump's administration has also declared war on fact-checking, arguing that it infringes on people's right to free speech.[26]

Attacks on truth and science are straight out of the authoritarian playbook. In the 1930s, Stalin elevated Trofem Lysenko's fringe theory of agriculture over the theory of Nikolai Vavilov, a leading geneticist famous for his seed bank to capture global diversity. For Lysenko and Stalin, a successful harvest wasn't about good genes but about how the crop was raised. Vavilov died in prison. The country starved. Mao's Cultural Revolution purged academics from universities, setting back

China's education system and research by decades and contributing to the starvation of some 30 million people. Hitler's attack on universities in Germany and Nazi-occupied Europe purged Jews and liberal scientists, sending many of them to America to develop the atomic bomb and supercharge American science.[27]

America's descent into the dark ages affects the entire world. So how will this book help you to navigate the present era? It is my hope that having read these stories, you will have imbibed something of the scientific method.

One thing should be clear: scientists are human beings. A PhD does not cancel overweening ambition, egotism, self-deception, cock-sureness or personality disorders. The scientific framework that keeps human foibles in check is that in the end, only the data counts. Science is self-correcting. But at any given moment, we all need to navigate a morass of conflicting claims.

That was pretty much the starting point for all the stories I've told here.

My guiding principles are:

Go to the experts. See what is being published in the leading journals and find a good plain-language summary. The world's two leading scientific publications, *Science* and *Nature*, both offer excellent free news reporting. *The New York Times* also does excellent science reporting. Check several sources.

Expert opinion seeks consensus – tough as that is to obtain among scientists. It is based on a convergence of evidence from different sources.[28] A scientist who is giving opinions outside their field of expertise is one to whom I give less weight. It's a red flag when scientists try to whip up an emotional rather than reasoned response, or if they have a predetermined opinion or conflict of interest. Whether this is intentional or unintentional, their finding is more likely to have a bias. A case in point was one of the influential but flawed studies that showed ivermectin was highly effective against Covid. One of the researchers, Tess Lawrie, headed a British ivermectin advocacy group.[29]

Another hallmark of scientific evidence is that it is not presented in black and white. It is measured. It comes with margins of error and estimates of effectiveness and risk.

This is the kind of language you'd find on an Australian website offering official advice on Covid vaccine efficacy and risks.[30] Once upon a time, it was also what you would have seen on an American government website.

No longer. Go to the White House website that previously provided information on Covid and you'll find a weird black and white graphic on the origins of the virus. One journalist described it as a fusion of airport-thriller fiction and wedding invite.[31]

Under the pretence of depoliticising the deep state, the Trump administration has *entirely,* brazenly, politicised it. The webpage is propaganda, full of factual errors, omissions and the rewriting of history. If science seeks to amass evidence from different sources, this new website simply omits all countervailing evidence: the incontrovertible fact that the virus spread from the wildlife market; that illegally traded raccoon dogs susceptible to the virus were there; that pangolins, the most trafficked wild animals in the world, carry one of the most similar viruses to SARS-CoV-2, as do bats found in Laos. Both the pangolins and bats lie on the path of a notorious and illegal animal trafficking route into southern China. The website also falsely states that there was only one introduction of the virus into humans; as we have seen, there were two, both traced to the market. It asserts that researchers at the Wuhan Institute of Virology "were sick with Covid-like symptoms in the fall of 2019", ignoring the previous director of national intelligence's findings that "some of the symptoms were not consistent with Covid-19".

Worst of all, it demonises Anthony Fauci, who served seven presidents over sixty years, advising them on outbreaks of HIV, killer flu, Ebola, Zika, anthrax and Covid. He and his family have faced death threats.

The history of authoritarian regimes tells us that when ideologues take over science, it does not end well.

Our health, our agriculture, our environmental safety, our ability to ameliorate and adapt to climate change, to regulate AI and to fight the next pandemic – all of these rely on our scientific system functioning as we know it must: through the generation of evidence, and rigorous testing of that evidence.

Acknowledgements

Writing a book takes a lot of crafting and polishing.

So very many friends and family members helped that process along by acting as sounding boards and giving me incisive feedback. Thank you particularly to Tamara Bruce, Kerry Bugalski, Natalie Bugalski, Shanit Borsky, Bella Hirshorn, Geoffrey Kempler, Jeremy Kenner, Harry Kestin, Lori Kravos, Sarah Luria, Helen Silver, Annette Smorgon, Cameron Tantau, Harrison Young and Audrey Zibelman.

And then there were those who read drafts and gave me vital feedback. Thank you, Katie Allen, Kristian Andersen, Lisa Lethin, Carlos Chaccour, David Copolov, Teresa Engelhard, Jock Findlay, Alan Finkel, Alex Finkel, Victor Finkel, Ron Finkel, Paul Glasziou, Mira Hetnal, Eddie Holmes, Brinley Hosking, Ros Hosking, Mark Kestin, Graeme Jackson, Ramona Koval, Michał Klincewicz, Richard Larkins, Ron Lazarovits, Allan Lew, Sharon Lewin, Bastien Llamas, John Matthewson, Christine McDonald, Michael Petraglia, Steve Redman, Ruth Rosen, Leanne Rowe and Tom Schwarz.

I also thank the many scientists I interviewed. Too many to list here, except for those who went above and beyond, giving up their time and energy over numerous sessions to answer my incessant questions. To Christof Koch, Liad Mudrik, Colin Pardoe, Michael Pitts, Bernard Schutz, Steven Smith, Michael Strevens and Chris Stringer – I thank you for that unstinting generosity of spirit that is the hallmark of a teacher.

As far as the philosophy of science, for me the most daunting of all the topics covered here, Sandy Boucher at the University of New England has been my unofficial "supervisor" these last five years and roundly critiqued each of my chapters. Thank you, Sandy, for your firm and unstinting guidance. Without you, this book would not exist.

Finally, for making this book readable, I feel extraordinarily fortunate to have had the input of many editors.

Chris Feik at La Trobe University Press inspired this book. He has a way of absorbing a ton of complex ideas and spitting out a gem. It was he who conceived of writing a book about how scientific proof works. Thanks for that, Chris, for all the catalytic conversations along the way, and for sending me this gem when I needed it: "muddled equals work equals progress".

Rachel Nowak, who has cast her expert eye over all my books, once again helped translate some of my early drafts from academic jargon to prose.

My husband, Alan Finkel, brought his exacting standards to the grammar, comprehension and scientific credibility of each chapter. At the eleventh hour, he also drew six beautiful pictures to capture 6000 of my words.

And finally, to La Trobe University Press editor Denise O'Dea. I never expected that on the final lap, when I was almost spent, I would have such a brilliant cheerleader. At times I felt Denise understood my book better than I did. Her incisive queries and determination to see every idea fully explained took the text to new heights.

List of Figures

Endnotes

Introduction

1 Aristotle, *Generation of Animals* III.10, translated by Arthur Platt, in *The Complete Works of Aristotle;* and Michael Strevens, "Why Aristotle Didn't Invent Modern Science", *Big Think*, 15 April 2021.

2 S. DiNardo et al., "Two-Tiered Regulation of Spatially Patterned *Engrailed* Gene Expression During *Drosophila* Embryogenesis", *Nature* 332, 1988, pp. 604–9.

3 Michael Strevens, *The Knowledge Machine: How an Unreasonable Idea Created Modern Science*, Penguin Books, 2022.

4 Ben Trubody, "Richard Feynman's Philosophy of Science", *Philosophy Now* 114, June–July 2016.

5 Hume, quoted in Strevens, *The Knowledge Machine*, p. 140.

6 Hume, *A Treatise of Human Nature*, 1739–40, Book 1, Part 3, Section 6, online at https://davidhume.org/texts/t/1/3/6.

7 Kenan Malik, "Einstein Got It: Philosophy and Science Go Hand in Hand", *The Guardian*, 24 February 2019.

8 C.D. Broad, quoted in the *Stanford Encyclopedia of Philosophy*, "Charlie Dunbar Broad", note 16, online at https://plato.stanford.edu/.

9 Hume, *A Treatise of Human Nature*, Book 1, Part 3, Section 6.

10 William Boyd, "Oh Vienna", *The Guardian*, 4 February 2012.

11 Karl Popper, "Knowledge and the Shaping of Reality" (1982), republished in *In Search of a Better World: Lectures and Essays from Thirty Years*, Routledge, 1994, p. 39.

12 Karl Popper, *Unended Quest: An Intellectual Autobiography*, 1976, p. 38, via Strevens, *The Knowledge Machine*, p. 296.

13 Popper, *In Search of a Better World*, p. 87.

14 "From Budapest: The Story of Imre Lakatos", *The Philosopher's Zone*, ABC Radio National, 23 September 2006.

15 John Horgan, "Profile: Reluctant Revolutionary: Thomas S. Kuhn Unleashed 'Paradigm' on the World", *Scientific American*, May 1991, pp. 40, 49.

16 David Klepper, "The CIA Believes Covid Likely Originated in a Lab but Has Low Confidence in Its Own Finding" *Los Angeles Times*, 25 January 2025; Katie Bo Lillis, "CIA Assesses Lab Leak Most Plausible Source of Covid-19, Though with Low Confidence", CNN, 25 January 2025.

17 Popper, "Knowledge and the Shaping of Reality" in *In Search of a Better World*, p. 5: "The philosophical relativism that hides behind the 'old and famous question' "What is truth"" may open the ways to evil things such as a propaganda of lies inciting men to hatred . . . Relativism is one of the many crimes committed by intellectuals. It is a betrayal of reason and of humanity. I suppose that the alleged relativity of truth defended by some philosophers results from mixing up the notions of truth and certainty; for in the case of certainty we may indeed speak of degrees of certainty . . . All of this is of great importance for jurisprudence and legal practice . . . it is the duty of every juror to try and to discover the objective truth to the best of his knowledge, and according to his conscience. But at the same time, he should be aware of his fallibility, of his uncertainty . . . This is also the case in science."

18 Charles Darwin, *The Origin of Species* (1859), Modern Library Edition, 1993, p. 646.

19 Roger Lewin, "Ancestors Worshipped", *Science* 224(4648), pp. 477–79.

20 Svante Pääbo, *Neanderthal Man*, Basic Books, 2014, p. 175.

21 Archibald Cochrane and Max Blythe, *One Man's Medicine: An Autobiography of Professor Archie Cochrane*, BMJ Books, 1989.

22 Iain Chalmers, "Trying to Do More Good Than Harm in Policy and Practice: The Role of Rigorous, Transparent, Up-to-Date Evaluations", *The ANNALS of the American Academy of Political and Social Science* 589(1), September 2003.

23 Statista, "Coronavirus (Covid-19) Deaths Worldwide per One Million Population as of 13 July 2022, by Country", online at https://www.statista.com/.

Chapter One

1 Michael Worobey, "I Called for More Research on the Covid 'Lab Leak Theory'. Here's What I Found Out", *Los Angeles Times*, 8 March 2023.

2 Michael Standaert and Eva Dou, "In Search for Coronavirus Origins, Hubei Caves and Wildlife Farms Draw New Scrutiny", *The Washington Post*, 11 October 2021.

3 Jon Cohen, "Why Many Scientists Say It's Unlikely That SARS-CoV-2 Originated from a Lab Leak", *Science*, 2 September 2021.

4 Smriti Mallapaty, "WHO Abandons Plans for Crucial Second Phase of COVID-Origins Investigation", *Nature*, 14 February 2023.

5 Florence Débarre, "What We Can and Cannot Learn from SARS-CoV-2 and Animals in Metagenomic Samples from the Huanan Market", *Virus Evolution* 10(1), 29 December 2023.

6 Jon Cohen, "It's Inexcusable: WHO Blasts China for Not Disclosing Potential Data on Covid-19's Origin", *Science*, 17 March 2023.

7 Kimmy Yam, "Trump Tweets About Coronavirus Using Term 'Chinese Virus'", NBC News, 17 March 2020.

8 Adam Kelsey, "WH Trade Adviser Pressed to Defend China Accusations", ABC News, 17 May 2020.

9 Michael Worobey, "Dissecting the Early Covid-19 Cases in Wuhan", *Science*, 3 December 2021, pp. 1202–4.

10 James D. Cherry and Paul Krogstad, "SARS: The First Pandemic of the 21st Century", *Pediatric Research* 56(1), July 2004, pp. 1–5.

11 World Bank, "Covid-19 to Plunge Global Economy into Worst Recession since World War II", 8 June 2020, online at https://www.worldbank.org/.

12 Mayo Clinic, "H1N1 (Swine Flu)", online at https://www.mayoclinic.org/.

13 Centers for Disease Control, "SARS Basic Fact Sheet", online at https://www.cdc.gov/.

14 Yong-Zhen Zhang and Edward C. Holmes, "A Genomic Perspective on the Origin and Emergence of SARS-CoV-2", *Cell* 181(2), 16 April 2020, pp. 223–227.

15 Pfizer Inc. and BioNTech SE, "Pfizer and BioNTech Announce Vaccine Candidate Against COVID-19 Achieved Success in First Interim Analysis from Phase 3 Study", 9 November 2020, online at https://www.pfizer.com/.

16 Institut Pasteur,"40 Years of HIV Discovery: The Virus Responsible for AIDS Is Identified on May 20, 1983", 15 May 2023, online at https://www.pasteur.fr/en/.

17 Helen Pearson, "Labs Crack Killer's Code", *Nature*, 15 April 2003.

18 Jeremy Farrar and Anjana Ahuja, *Spike: The Virus vs the People*, Profile Books, 2022, p. 16.

19 David Quammen, *Breathless: The Scientific Race to Defeat a Deadly Virus*, Vintage, 2024, p. 17.

20 David Cyranoski, "Zhang Yongzhen: Genome Sharer", *Nature*, 15 December 2020.

21 Farrar and Ahuja, *Spike*, pp. 19–26; Smriti Mallapaty, "Chinese Virologist Who Was First to Share Covid-19 Genome Sleeps on Street after Lab Shuts", *Nature*, 1 May 2024.

22 Cyranoski, "Zhang Yongzhen: Genome Sharer".

23 Young Chan Kim, Barbara Dema and Arturo Reyes-Sandoval, "Covid-19 Vaciines: Breaking Record Times to First-in-Human Trials", *npj Vaccines* 5(34), 2020.

24 Worobey, "Dissecting the Early Covid-19 Cases in Wuhan"; Carl Zimmer, Benjamin Mueller and Chris Buckley, "First Known Covid Case Was Vendor at Wuhan Market, Scientists Say", *The New York Times*, 18 November 2021.

25 Peter Li, "Reopening the Trade after SARS: China's Wildlife Industry and the Fateful Policy Reversal", *Environmental Policy and Law* 50, 2020, pp. 1–17.

26 Moira Chan-Yeung and Rui-Heng Xu, "SARS: Epidemiology", *Respirology* 8(S1), November 2003, pp. S9–S14.

27 Cherry and Krogstad, "SARS: The First Pandemic of the 21st Century"; Y. Guan et al., "Isolation and Characterisation of Viruses Related to SARS Coronavirus from Animals in Southern China", *Science* 302(5643), 10 October 2003, pp. 276–78.

28 Ian Sample and John Gittings, "In China the Civet Cat Is a Delicacy – and May Have Caused SARS", *The Guardian*, 24 May 2003.

29 K.R. Drew, "Deer Farming in Asia", Proceedings of the 2nd World Deer Farming Congress, 1998, online at https://deernz.org.nz; Michael Standaert, "Coronavirus Closures Reveal Vast Scale of China's Secretive Wildlife Farm Industry", *The Guardian*, 25 February 2020.

30 Peter Li, Zoom interview, 16 September 2023.

31 Y. Guan et al., "Isolation and Characterisation of Viruses Related to SARS Coronavirus".

32 Except for three civets from farms in Hubei province, farmed civet populations were largely free of the virus and antibodies. So too with surveys of wild civets. They may have suffered an outbreak, but the infection burned out, perhaps because infected animals died.

33 Peter Li, "China's Wildlife Market, Trade and Consumption: Debunking Misperception and Falsehood", Animal Welfare Observatory, 26 May 2020, online at https://animal-welfareobservatory.org.

34 World Health Organization, "Hendra Virus Infection", online at https://www.who.int/health-topics.

35 P.L. Young, "Serologic Evidence for the Presence in Pteropus Bats of a Paramyxovirus Related to Equine Morbillivirus", *Emerging Infectious Diseases* 2(3), July–September 1996, pp. 239–40; Olivia Willis, "The Deadly Disease Outbreak in Sleepy Suburbia", *Patient Zero,* ABC Radio National, 5 July 2021.

36 Email from Linfa Wang, 5 September 2024.

37 Linfa Wang et al., "From Hendra to Wuhan: What Has Been Learned in Responding to Emerging Zoonotic Viruses", *Lancet* 395(10224), pp. e33–34, 22 February 2020.

38 Dennis Normile, "Researchers Tie Deadly SARS Virus to Bats", *Science* 309(5744), 30 September 2005, pp. 2154–55.

39 Xing-Yi Ge et al., "Isolation and Characterization of a Bat SARS-like Coronavirus that Uses the ACE2 Receptor", *Nature* 503, 2013, pp. 535–38.

40 Amanda Wilson et al., "A Thematic Analysis of Experiences of HIV Risks Among Female Sex Workers in the Yunnan–Vietnam Chinese Border Region", *BMC Women's Health* 21(7), 2021.

41 Xing-Yi Ge et al., "Isolation and Characterization of a Bat SARS-like Coronavirus".

42 Ben Hu et al., "Discovery of a Rich Gene Pool of Bat SARS-Related Coronaviruses Provides New Insights into the Origin of SARS Coronavirus", *PLOS Pathogens*, 30 November 2017.

43 Ecohealth Alliance, "Living Safely with Bats", online at https://www.ecohealthalliance.org.

44 Hu et al., "Discovery of a Rich Gene Pool of Bat SARS-Related Coronaviruses".

45 Hu et al., "Discovery of a Rich Gene Pool of Bat SARS-Related Coronaviruses".

46 David Cyranoski, "Inside the Chinese Lab Poised to Study World's Most Dangerous Pathogens", *Nature* 542, 2017, pp. 399–400.

47 Robert Walgate, "SARS Escaped Beijing Lab Twice", *Genome Biology* 4, 2004. doi: 10.1186/gb-spotlight-20040427-03.

48 Martin Enserink, "Tiptoeing Around Pandora's Box", *Science* 305(5684), pp. 594–95.

49 World Health Organization (Western Pacific Region), "Avian Influenza Weekly Update 983", 31 January 2025, online at https://www.who.int/.

50 Martin Enserink, "Controversial Studies Give a Deadly Flu Virus Wings", *Science* 334(6060), 2 December 2011, pp. 1192–93.

51 Ed Yong, "The Risks and Benefits of Publishing Mutant Flu Studies", *Nature*, 2 March 2012; Marc Lipsitch, "Why Do Exceptionally Dangerous Gain-of-Function Experiments in Influenza?", Methods in Molecular Biology, 2018, doi: 10.1007/978-1-4939-8678-1_29.

52 Lisa Schnirring, "H5N1 Avian Flu Viruses Can Persist on Milking Equipment Surfaces", *CIDRAP*, 24 May 2024.

53 Declan Butler, "Engineered Bat Virus Stirs Debate Over Risky Research", *Nature*, 2015.

54 Jocelyn Kaiser, "NIH Lifts 3-Year Ban on Funding Risky Virus Studies", *Science*, 19 December 2017.

55 David Cyranoski, "Inside the Chinese Lab Poised to Study World's Most Dangerous Pathogens".

56 Jon Cohen, "Anywhere but Here", *Science*, 18 August 2022.

57 Xinhua, "China's Legislature Adopts Decision on Banning Illegal Trade, Consumption of Wildlife", *Xinhuanet*, 24 February 2020; Dennis Normille, "China Is Cracking Down on Its Wildlife Trade. Is it Enough?", *Science Insider*, 29 March 2023.

58 Michael Standaert and Eva Dou, "In Search for Coronavirus Origins, Hubei Caves and Wildlife Farms Draw New Scrutiny", *The Washington Post*, 11 October 2021.

59 Jon Cohen, "Wuhan Coronavirus Hunter Shi Zhengli Speaksk Out", *Science*, 24 July 2020; Jane Qiu, "How China's 'Bat Woman' Hunted Down Viruses from SARS to the New Coronavirus", *Scientific American*, 1 June 2020.

60 Jon Cohen, "Call of the Wild", *Science*, 2 September 2021.

61 P. Zhou, X.L. Yang, X.G. Wang et al., "A Pneumonia Outbreak Associated with a New Coronavirus of Probable Bat Origin", *Nature* 579, 2020, pp. 270–73 (2020).

62 David Quammen, *Breathless*, Penguin, 2022, p. 82.

63 Kristian Andersen, Zoom interview, 5 April 2024 unless otherwise specified.

64 Quammen, *Breathless*, p. 79.

65 Shayla Love, "Once and for All, the New Coronavirus Was Not Made in a Lab", *Vice*, 20 March 2020.

66 Farrar and Ahuja, *Spike*, p. 65.

67 Cohen, "Call of the Wild".

68 Farrar and Ahuja, *Spike*, p. 69.

69 K.G. Andersen, A. Rambaut, W.I. Lipkin et al., "The Proximal Origin of SARS-CoV-2", *Nature Medicine* 26, 2020, pp. 450–52.

70 Danile Wrapp et al., "Cryo-EM Structure of the 2019-nCoV Spike in the Prefusion Conformation", *Science* 367(6483), 19 February 2020, pp. 1260–63.

71 Jane Qiu, "Meet the Scientist at the Center of the Covid Lab Leak Controversy", *MIT Technology Review*, 9 February 2022.

72 M.F. Boni, P. Lemey, X. Jiang et al., "Evolutionary Origins of the SARS-CoV-2 Sarbecovirus Lineage Responsible for the COVID-19 Pandemic", *Nature Microbiology* 5, 2020, pp. 1408–17.

73 Kristian Andersen, email correspondence, 2 November 2024.

74 Gary Whittaker, telephone interview, 1 September 2023, and email correspondence; Jing Lu, "Identification of a Common Deletion in the Spike Protein of SARS CoV-2", Virological.org, March 2020.

75 Yiran Wu and Suwen Zhao, "Furin Cleavage Sites Naturally Occur in Coronaviruses", *Stem Cell Research*, December 2020, doi: 10.1016/j.scr.2020.102115.

76 K.G. Andersen, A. Rambaut, W.I. Lipkin et al., "The Proximal Origin of SARS-CoV-2".

77 James Gorman and Carl Zimmer, "Scientist Opens Up About His Early Email to Fauci on Virus Origins", *The New York Times*, 14 June 2021.

78 Among the ranks of the conspiracists is economist Jeffrey Sachs, director of the Center for Sustainable Development at Columbia University. His hypothesis is that Ralph Baric constructed SARS-CoV-2 in his lab in North Carolina and sent it to Zhengli Shi in Wuhan. See "Origin of Covid and Whistleblowing with Jeffrey Sachs", Judge Napolitano, https://www.youtube.com/watch?v=nyqoBiq7cz8.

79 Nicholas Wade, "The Origin of Covid: Did People or Nature Open Pandora's Box at Wuhan?", *Bulletin of the Atomic Scientists*, 5 May 2021.

80 Sharri Markson, "Covid Cover-Up: Wuhan Suspicions, Anthony Fauci and How the Science Was Silenced", *The Australian*, 28 July 2023.

81 Jon Cohen, "Science Takes Back Seat to Politics in First House Hearing on Origin of Covid-19 Pandemic", *Science*, 8 March 2023.

82 Daniel Engber, "The Republican Lab-Leak Circus Makes One Important Point", *The Atlantic*, 12 July 2023.

83 Jon Cohen, "Politicians, Scientists Spar Over Alleged NIH Cover-Up Using Covid-19 Origin Paper", *Science*, 11 July 2023.

84 Jocelyn Kaiser, "'Lab-Leak' Proponents at Rutgers Accused of Defaming and Intimidating Covid-19 Origin Researchers", *Science*, 15 March 2024.

85 Jon Cohen, "Anthony Fauci on Becoming the 'Devil' and a Warning for His Successor", *Science*, 25 September 2023.

86 Jane Qiu, "This Scientist Now Believes Covid Started in Wuhan's Wet Market. Here's Why", *MIT Technology Review*, 23 November 2021.

87 Quammen, *Breathless*, p. 136.

88 Michael Worobey et al., "Contaminated Polio Vaccine Theory Refuted", *Nature* 428, 2004, p. 820.

89 World Health Organization, *WHO-Convened Global Study of Origins of SARS-CoV-2: China Part*, 30 March 2021.

90 Jesse D. Bloom et al., "Investigate the Origins of Covid-19", *Science* 372(6543), 14 May 2021, p. 694.

91 Shing Hei Zhan, Benjamin E. Deverman and Yujia Alina Chan, "SARS-CoV-2 Is Well Adapted for Humans. What Does This Mean for Re-Emergence?", bioRxiv, 2 May 2020, https://doi.org/10.1101/2020.05.01.073262.

92 Quammen, *Breathless*, p. 124.

93 Zimmer, Mueller and Buckley, "First Known Covid Case Was Vendor at Wuhan Market, Scientist Says".

94 Worobey, "Dissecting the Early Covid-19 Cases in Wuhan".

95 Michael Worobey, Zoom interview, 8 September 2023.

96 Michael Worobey et al., "The Huanan Seafood Wholesale Market in Wuhan Was the Early Epicenter of the Covid-19 Pandemic", *Science* 377(6609), 26 July 2022, pp. 951–59.

97 Cohen, "It's Inexcusable"; Florence Débarre, "What We Can and Cannot Learn from SARS-CoV-2 and Animals in Metagenomic Samples from the Huanan Market", *Virus Evolution* 10(1), 2024.

98 Sofía Jijón, Peter Czuppon, François Blanquart and Florence Débarre, "Using Early Detection Data to Estimate the Date of Emergence of an Epidemic Outbreak", *PLOS Computational Biology*, 8 March 2024; James Gallagher, "Genetic Ghosts Suggest Covid's Origins", BBC News, 20 September 2024; W.J. Liu, P. Liu, W. Lei et al., "Surveillance of SARS-CoV-2 at the Huanan Seafood Market", *Nature* 631, 2024, pp. 402–8; Alexander Crits-Christoph et al., "Genetic Tracing of Market Wildlife and Viruses at the Epicenter of the Covid-19 Pandemic", *Cell* 187(19), 19 September 2024.

99 Karen Kaplan, "New Genetic Research Points to Wuhan Animal Market as Origin of Covid Pandemic, Study Says", *Los Angeles Times*, 21 September 2024.

100 J.E. Pekar et al., "The Molecular Epidemiology of Multiple Zoonotic Origins of SARS-CoV-2.", *Science* 377, 2022, pp. 960–66.

101 Jijón et al., "Using Early Detection Data to Estimate the Date of Emergence of an Epidemic Outbreak".

102 Daoyu Zhang et al., "DRASTIC: An Analysis of Project DEFUSE", September 2021, https://10.13140/RG.2.2.12961.89442; Jamie Palmer, "The Lab-Leak Illusion", *Quillette*, 19 August 2023.

103 DEFUSE proposal, online at https://www.documentcloud.org/documents/21066966-defuse-proposalOur.

104 Sharon Lerner and Maia Hibbett, "Leaked Grant Propposal Details High-Risk Coronavirus Research", *The Intercept*, 23 September 2021.

105 Josh Hawwley and Mike Braun, "Hawley and Braun Demand Biden Implement Covid Origins Act", 15 June 2023, online at https://www.hawley.senate.gov/.

106 Office of the Director of National Intelligence, "Potential Links Between the Wuhan Institute of Virology and the Origin of the Covid-19 Pandemic", June 2023, online at https://www.dni.gov/.

107 Sharri Markson, "Edward Holmes Claims 'Bad Memory' for Not Declaring Writing a Paper with Wuhan Scientist", *The Australian*, 31 July 2023.

108 Cohen, "Call of the Wild"; James C. Alwine et al., "A Critical Analysis of the Evidence for the SARS-CoV-2 Origin Hypothesis", *mSphere* 8(2), 20 April 2023.

109 Stuart Turville, telephone interview, 2 October 2023.

110 Linfa Wang, Zoom interview, 4 August 2023.

111 Eric Donaldson, Amy Sims and Ralph Baric, "Systematic Assembly and Genetic Manipulation of the Mouse Hepatitis Virus A59 Genome", *Methods of Molecular Biology* 454, 2008, pp. 293–315.

112 Committee on Oversight and Accountability, Select Subcommittee on the Coronavirus Pandemic, US House of Representatives. Interview of Ralph S. Baric, 22 January 2024, p. 63, online at https://oversight.house.gov/wp-content/uploads/2024/04/Baric-TI-Transcript.pdf

113 "thirty letters per year": estimate by Stuart Turville using bioinformatic tools from Nextstrain, https://nextstrain.org/.

114 S. Temmam et al., "Bat Coronaviruses Related to SARS-CoV-2 and Infectious for Human Cells", *Nature* 604, 2022, pp. 330–36.

115 J. Chen et al., "A Bat MERS-like Coronavirus Circulates in Pangolins and Utilizes Human DPP4 and Host Proteases for Cell Entry", *Cell* 186(4), 16 February 20223, pp. 850–63.

116 M.A. Peña-Hernández et al., "SARS-CoV-2-Related Bat Viruses Evade Human Intrinsic Immunity but Lack Efficient Transmission Capacity", *Nature Microbiology* 9, 2024, pp. 2038–50. The viruses lack the FCS, suggesting it is important for transmission efficiency.

117 Xiao Xiao et al., "Animal Sales from Wuhan Wet Markets Immediately Prior to the Covid-19 Pandemic", *Scientific Reports* 11, 2021.

118 Centers for Disease Control, "Animals and Covid-19", online at www.cdc.gov; Jon Cohen, "Virus Gone Wild", *Science*, 29 April 2024.

119 Ning Wang et al., "Serological Evidence of Bat SARS-Related Coronavirus Infection in Humans, China", *Virologica Sinica* 33(1), 2 March 2018, pp. 104–7.

120 Peter Li, Zoom interview, 16 September 2023.

121 Alina Chan, "Why the Pandemic Probably Started in a Lab, in Five Key Points", *The New York Times*, 3 June 2024.

122 James Alwine et al., "The Harms of Promoting the Lab Leak Hypothesis for SARS-CoV-2 Origins without Evidence", *Journal of Virology* 98(9), 1 August 2024.

123 David Leonhardt, "Two Covid Theories", *The New York Times*, 14 June 2024.

Chapter Two

1 Richard Klein, "Darwin and the Recent African Origin of Modern Humans", *PNAS* 106(38), 22 September 2009, pp. 16007–9; John van Wyhe and Peter C. Kjærgaard, "Going the Whole Orang: Darwin, Wallace and the Natural History of Orangutans", *Studies in History and Philosophy of Science Part C: Biological and Biomedical Sciences*, vol. 51, June 2015, pp. 53–63; John James Wilson, "Saving the 'Land of the Orangutan'", National Museums Liverpool, online at https://www.liverpoolmuseums.org.uk.

2 All dates should be viewed as "estimates". They often change as new techniques are used.

3 David Hume, *Treatise of Human Nature*, 1739–40, online at https://davidhume.org/texts/t/full.

4 David Reich, *Who We Are and How We Got Here*, Oxford University Press, 2018, p. 91.

5 Fernando Racimo et al., "Evidence for Archaic Adaptive Introgression in Humans", *Nature Reviews Genetics* 16, 2015, pp. 359–71.

6 Hominin is a term for the group consisting of modern humans and their immediate ancestors that emerged since we split from our common ancestor with chimpanzees around 7 million years ago.

7 Jim Allen, "The Curious History of the Talgai Skull", *Bulletin of the History of Archaeology* 20(2), 2010, pp. 4–12.

8 P. Cunningham, *Two Years in New South Wales* (1827), Libraries Board of South Australia facsimile edition, 1966. Quoted in Jim Allen, "The Curious History of the Talgai Skull".

9 Joseph Birdsell, "Preliminary Data on the Trihybrid Origin of the Australian Aborigines", *Archaeology and Physical Anthropology in Oceania* 2(2), 1967, pp. 100–155.

10 Michael Westaway and Peter Hiscock, "The Extinction of Rigour: A Comment on the Windschuttle and Gillin Article 'The Extinction of Pygmies'", *Aboriginal History* 29, 2005, p. 142.

11 Elizabeth Culotta and Ann Gibbons, "Almost All Living People Outside of Africa Trace Back to a Single Migration More than 50,000 Years Ago", *Science*, 21 September 2016.

12 Online variorum of Darwin's *Origin of Species*: fourth British edition (1866), p. 155, online at darwin-online.org.uk.

13 Steven Rose, "Darwin, Race and Gender", *EMBO Reports* 10, April 2009, pp. 297–98.

14 Darwin's 1857 letter to Alfred Russel Wallace, quoted by Bruce Alberts, "Understanding Human Origins", *Science* 326(5949), 2 October 2009, p. 17.

15 Darwin, *On the Origin of Species*, p. 647.

16 Charles Steffen, "The Man Behind the Eponym: Thomas Henry Huxley", *The American Journal of Dermatopathology* 24(1), February 2002, pp. 82–84; Thomas Henry Huxley, "On the Anatomy and the Affinities of the Family of the Medusae", *Philosophical Transactions* 139, 1849; Glenn Branch, "How Stupid Not to Have Thought of That", National Center for Science Education, 13 January 2015, online at https://ncse.ngo/.

17 Jonathan Smith, "The Huxley–Wilberforce 'Debate' on Evolution, 30 June 1860", BRANCH, online at https://branchcollective.org/.

18 Janet Browne, *Charles Darwin, Vol. 2: The Power of Place*, Jonathan Cape, 2002.

19 Klein, "Darwin and the Recent African Origin of Modern Humans".

20 Giacomo Bernardi, "Following Alfred Russel Wallace's Footsteps to Borneo, Where He Penned His Pioneering Evolution Paper", *The Conversation*, 15 August 2018.

21 University of California Berkeley, *The History of Evolutionary Thought*, online at https://evolution.berkeley.edu/.

22 L. Leakey, P. Tobias and J. Napier, "A New Species of the Genus Home from Olduvai Gorge", *Nature* 202, 1964, pp. 7–9.

23 Chris Stringer, interviews and email correspondence, 23 December 2020 through 21 February 2025.

24 Kevin Blake, "On the Origins of 'The March of Progress'", Washington University ProSPER, 17 December 2018, online at https://sites.wustl.edu/prosper/.

25 Smithsonian Museum of Natural History, "Homo neanderthalensis", online at https://humanorigins.si.edu/.

26 Kilbee Brittain, "Book Shelf: The Strange Case of the Rickety Cossack", The Leakey Foundation, 25 January 2016, online at https://leakeyfoundation.org.

27 Geologist William King suggested the name. Smithsonian Museum, "Homo neanderthalensis", https://humanorigins.si.edu/.

28 Chris Stringer, personal communication, 21 February 2025.

29 James Shreeve, "Erectus Rising", *Discover Magazine*, 1 September 1994; Pat Shipman, *The Man Who Found the Missing Link*, Simon and Schuster, 2001, pp. 20, 37.

30 Franz Weidenrich, *Apes, Giants and Man*, University of Chicago Press, 1946, p. 26.

31 Smithsonian Museum of Natural History, "Trinil 2", online at https://humanorigins.si.edu/.

32 Yahdi Zaim et al., "New 1.5 Million-Year-Old Homo Erectus Maxilla from Sangiran (Central Java, Indonesia", Journal of *Human Evolution* 61(4), October 2011, pp. 363–76.

33 Y. Rizal et al., "Last Appearance of *Homo Erectus* at Ngangdong, Java, 117,000–108,000 Years Ago", *Nature* 577, 2020, pp. 381–85.

34 Noel T. Boaz and Russell L. Ciochon, *Dragon Bone Hill: An Ice-Age Saga of Homo Erectus*, Oxford University Press, pp. 3–4.

35 Davidson Black, "Tertiary Man in Asia: The Chou Kou Tien Discovery", *Nature* 118, 1926, pp. 733–34.

36 Boaz and Ciochon, *Dragon Bone Hill*, p. 19.

37 Russell Ciochon and E. Arthur Bettis III, "Asian *Homo Erectus* Converges in Time", *Nature* 458, 2009, pp. 153–54; G. Shen et al,. "Age of Zhoukoudian *Homo Erectus* Determined with 26Al/10Be Burial Dating", *Nature* 458, 2009, 198–200.

38 Boaz and Ciochon, *Dragon Bone Hill*, pp. 29–30.

39 Smithsonian Museum of Natural History, "Homo Heidelbergensis", online at https://humanorigins.si.edu/.

40 Weidenreich, *Apes, Giants and Man*, University of Chicago Press, 1946, p. 84.

41 R. Kimura, "A Common Variation in *EDAR* Is a Genetic Determinant of Shovel-Shaped Incisors", *Cell* 85(4), October 2009, pp. 538–35; Yana G. Kamberov et al., "Modeling Recent Human Evolution in Mice by Expression of a Selected EDAR Variant", *Cell* 152(4), February 2013, pp. 691–702.

42 Milford Wolpoff and Rachel Caspari, *Race and Human Evolution*, Simon and Schuster, 1997, p. 177; "Dr Franz Weidenreich on Human Evolution, *Nature* 162, 1948, p. 484.

43 Michael Bird et al., "Palaeogeography and Voyage Modeling Indicates Early Human Colonization of Australia Was Likely from Timor-Roti", *Quaternary Science Reviews*, 2018; and email communication with Michael Bird, 4 February 2025.

44 "The Talgai Skull", *Nature* 96, 1915, pp. 52–53.

45 J. Allen, The Curious History of the Talgai Skull", *Bulletin of the History of Archaeology* 20(2), 2010, pp. 4–12.

46 A.D. Rothman, "Cohuna Skull Called Most Archaic Known; Australian Museum Director Asserts It Is From Direct Human Line - Other Scientists Unconvinced by Facts So Far Presented - May Be Early Bush Man", *The New York Times*, 7 November 1926.

47 M.H. Monroe, "Australia: The Land Where Time Began", 2011, https://austhrutime.com/cohuna.htm.

48 Franz Weidenreich, "Some Problems Dealing with Ancient Man", *American Anthropologist, New Series* 42(3.1), July–September 1940, pp. 375–83.

49 Desmond, "Darwin the Abolitionist".

50 Boaz and Ciochon, *Dragon Bone Hill*, pp. 29–30.

51 R. Caspari and M.H. Wolpoff, "Weidenreich, Coon, and Multiregional Evolution", *Human Evolution* 11 1996, 261–68.

52 Adrian Desmond, "Darwin the Abolitionist", *Prospect*, February 2009.

53 Boaz and Ciochon, *Dragon Bone Hill*, p. 71.

54 Desmond, "Darwin the Abolitionist".

55 Robin McKie, "Cheddar Man Changes the Way We Think About Our Ancestors", *The Guardian*, 11 February 2018.

56 Reich, *Who We Are and How We Got Here*, p. 82. See also Reich's defence of his book: "How to Talk About 'Race' and Genetics", *The New York Times*, 30 March 2018.

57 Templeton, "Biological Races in Humans".

58 N.W.G. Macintosh, "The Cohuna Cranium: History and Coommentary from November 1925 to November 1951", *The Australian Journal of Anthropology*, February 2009.

59 A. Thorne and P. Macumber, "Discoveries of Late Pleistocene Man at Kow Swamp, Australia", *Nature* 238, 1972, pp. 316–19.

60 Billy Griffiths, *Deep Time Dreaming*, Black Inc., 2018, p. 117.

61 "Mungo Archaeology", online at visitmungo.com.au.

62 Griffiths, *Deep Time Dreaming*, pp. 122–23.

63 Griffiths, Deep Time Dreaming, p. 131; "Mungo Lady", National Museum Australia, online at www.nma.gov.au; E.E. Wreschner et al., "Red Ochre and Human Evolution", *Current Anthropology* 21(5), 1980.

64 Milford Wolpoff, telephone interview, February 2021.

65 Milford Wolpoff and Rachel Caspari, *Race and Human Evolution*, Simon and Schuster, 1997, p. 28.

66 Milford Wolpoff, telephone interview, 13 January 2021; Wolpoff and Caspari, *Race and Human Evolution*.

67 M.H. Wolpoff et al., "Modern Homo Sapiens Origins: A General Theory of Hominid Evolution Involving the Fossil Evidence from East Asia", in *The Origins of Modern Humans*, Liss, 1984, pp. 411–83; see also Thorne and Wolpoff, "Regional Continuity in Australian Pleistocene Hominid Evolution", *American Journal of Physical Anthropology* 55(3), July 1981, pp. 337–34.

68 Wolpoff and Caspari, *Race and Human Evolution*, p. 257.

69 I. McDougall, F. Brown and J. Fleable, "Stratigraphic Placement and Age of Modern Humans from Kibish, Ethiopia", *Nature* 433, 2005, pp. 733–36.

70 C.B. Stringer, "Population Relationships of Later Pleistocene Hominids: A Multivariate Study of Available Crania", *Journal of Archaeological Sciences* 1(4), December 1974, pp. 317–42.

71 S.J. Gould, *The Structure of Evolutionary Theory*, Harvard University Press, 2002, quoted in Alan Templeton, "Genetics and Recent Human Evolution", *Evolution* 61(7), July 2007, pp. 1507–19.

72 Peter Brown, telephone interview, June 2021.

73 Peter Brown, "Nacurrie 1, Mark of Ancient Java, or a Caring Mother's Hands, in Terminal Pleistocene Australia?, *Journal of Human Evolution* 59(2), August 2010, pp. 168–87.

74 Michael C. Westaway and Colin P. Groves, "The Mark of Ancient Java is on None of Them", *Archaeology in Oceania* 44(2), July 2009, pp. 84–95.

75 Colin Groves, "Alan Gordon Thorne, 139–2012", T*he Australian Academy of the Humanities 2012 Annual Report*, pp. 43–44.

76 Ian Gilligan and David Bulbeck, "Environment and Morphology in Australian Aborigines: A Reanalysis of the Birdsell Database", *American Journal of Biological Anthropology* 134(1), September 2007, pp. 75–91.

77 Colin Pardoe, interviews and email correspondence, June–July 2021.

78 Rebecca L. Cann, Mark Stoneking and Allan C. Wilson, "Mitochondrial DNA and Human Evolution", *Nature* 325, 1987, pp. 31–36.

79 Sam Wood, "Languages Have Families, Too: A Look at the Indo-European Language Family", Babel.com, 22 February 2020.

80 C.B. Stringer and P. Andrews, "Genetic and Fossil Evidence for the Origin of Modern Humans", *Science* 239(4845), March 1988, pp. 1263–68.

81 M.H. Wolpoff, J. Hawks and R. Caspari, "Multiregional, Not Multiple Origins", *American Journal of Physical Anthropology* 112(1), May 2000, pp. 129–36.

82 Patrick D. Evans et al., "Evidence That the Adaptive Allele of the Brain Size Gene Microcephalin Introgressed into Homo Sapiens from an Archaic Homo Lineage", *Procedings of the National Academy of Sciences* 103(48), 7 November 2006.

83 Marcia Barinaga, "'African Eve' Backers Beat a Retreat", *Science* 255(5045), 7 February 1992, pp. 687–87.

84 Alan Templeton, Zoom interview, 12 February 2021, and email correspondence; A.R. Templeton, "Nested Clade Analyses of Phylogeographic Data: Testing Hypotheses about Genetic Population History", *Molecular Ecology* 7(4), April 1998, pp. 381–97; Alan Templeton, "Bilogical Races in Humans"; Alan Templeton, *Human Population Genetics and Genomics*, Elsevier, 2018, Chapter 6.

85 Mark Stoneking, Zoom interview, 25 February 2021. A co-author of Allan Wilson's seminal 1987 paper, he provided an edifying overview of the history of this period. See also: Anna Di Rienzo and Allan C. Wilson, "Branching Pattern in the Evolutionary Tree for Human Mitochondrial DNA", *PNAS* 88(5), 1 March 1991, pp. 1597–1601; Carl Zimmer, "A Single Migration from Africa Populated the World, Studies Find", *The New York Times*, 22 September 2016.

86 Pääbo, *Neanderthal Man*, Chapter 2.

87 Svante Pääbo, "Molecular Cloning of Ancient Egyptian Mummy DNA", *Nature* 314 (1985), pp. 644–45.

88 R. Thomas et al., "DNA Phylogeny of the Extinct Marsupial Wolf", *Nature* 340, 1989, pp. 465–67.

89 Pääbo, *Neanderthal Man*, p. 51.

90 O. Handt et al., "Molecular Genetic Analyses of the Tyrolean Ice Man", *Science* 264(5166), June 1994, pp. 1775–78.

91 Matthias Krings et al., "Neandertal DNA Sequences and the Origin of Modern Humans", *Cell* 90(1), July 1997, pp. 19–30.

92 R.E. Green, "A Draft Sequence of the Neandertal Genome", *Science* 328(5979), 7 May 2010, pp. 710–22.

93 Pääbo, *Neanderthal Man*, p. 188.

94 Tomislav Maricic et al., "A Recent Evolutionary Change Affects a Regulatory Element in the Human FOXP2 Gene", *Molecular Biology and Evolution*, 30(4), April 2013, pp. 844–52; Rockefeller University, "Researchers Identify a Gene Linked to the Evolution of Spoken Language", 18 February 2025, online at www.rockefeller.edu.

95 L. Skov et al., "The Nature of Neanderthal Introgression Revealed by 27,566 Icelandic Genomes", *Nature* 582, 2020, pp. 78–83; Janet Kelso, email communication, 6 June 2021.

96 Freda Kreier, "Neanderthals and Humans Interbred More Recently Than Scientists Thought", *Nature*, 12 December 2024.

97 Harvard Medical School, "Neanderthals' Genetic Legacy: Humans Inherited Variants Affecting Disease Risk, Infertility, Skin and Hair Characteristics", *Science Daily*, 29 January 2014.

98 Yang Luo, "Neanderthal DNA Highlights Complexity of COVID Risk Factors", *Nature*, 26 October 2020.

99 D. Reich et al., "Genetic History of an Archaic Hominin Group from Denisova Cave in Siberia", *Nature* 468(7327), December 2010, pp. 1053–60.

100 Ann Gibbons, "Who Were the Denisovans?", *Science* 333(6046), 26 August 2011.

101 Matthew Waren, "Mum's a Neanderthal, Dad's a Denisovan: First Discovery of an Ancient-Human Hybrid", *Nature*, 22 August 2018.

102 Gibbons, "Who Were the Denisovans?"

103 Gibbons, "Who Were the Denisovans?"

104 As attendsted by occupation at the Madjedbebe rock shelter: C. Clarkson et al., Human Occupation of Northern Australia by 65,000 Years Ago", *Nature* 547, 2017, pp. 306–10.

105 David Reich et al., "Denisova Admixture and the First Modern Human Dispersals into Southeast Asia and Oceania", *The American Journal of Human Genetics* 89(4), pp. 516–28.

106 10,000 years ago it included Java and other islands and was known as Sunda.

107 Annie Melchor, "Indigenous Filipino Group Has Highest Known Denisovan Ancestry", *The Scientist*, 14 August 2021.

108 Guy Jacobs et al., "Multiple Deeply Divergent Denisovan Ancestries in Papuans", *Cell* 177(4), 2 May 2019.

109 A common paraphrasing of what Darwin wrote in *Origin of Species*, variorum edition, p. 47, via Richard L. Pyle, "Towards a Global Names Architecture: The Future of Indexing Scientific Names", *ZooKeys* 550, January 2016, pp. 261–81.

110 Ewen Callaway, "Siberia's Ancient Ghost Clan Starts to Surrender Its Secrets", *Nature*, 27 February 2019.

111 Emilia Huerta-Sanchez, "Altitude Adaptation in Tibetans Caused by Introgression of Denisovan-like DNA", *Nature* 512(7513), August 2014, pp. 194–97.

112 Jean-Jacques Hublin, "How We Found an Elusive Hominin in China", Sapiens.org, 1 May 2019.

113 Ann Gibbons, "Ancient Jaw Gives Elusive Denisovans a Face", *Science* 364(6439), pp. 418–19.

114 Ann Gibbons, "DNA Tracks Mysterious Denisovans to Chinese Cave, Just Before Modern Humans Arrived Nearby", *Science*, 29 October 2020.

115 J.C. Teixeira et al., "Widespread Denisovan Ancestry in Island Southeast Asia but No Evidence of Substantial Super-Archaic Hominin Admixture", *Nature Ecology and Evolution* 5, 2021, pp. 616–24; Elizabeth Finkel, "Tightening the Dragnet on Denisovans", *Cosmos*, 23 March 2021. But other researchers do report finding traces of "superarchaic" DNA in European populations: PLOS, "DNA from an Ancient, Unidentified Ancestor Was Passed Down to Humans Living Today", *Science Daily*, 6 August 2020.

116 Roberto Rozzi et al., "Dwarfism and Gigantism Drive Human-Mediated Extinctions on Islands", *Science* 379(6636), 9 March 2023, pp. 1054–59.

117 Kris Helgen, telephone interview, January 2021.

118 João Teixeira, telephone interviews and email correspondence, January 2021.

119 Ann Gibbons, "Meet the Frail, Small-Brained People Who First Trekked Out of Africa", *Science*, 22 November 2016.

120 Heather Pringle, "Quest for Fire Began Earlier Than Thought", *Science*, 2 April 2012.

121 John Hawks, telephone interview, 15 February 2021.

122 Elizabeth Finkel, "Cue the Laos Cobra Cave Tooth, First Clue in the Denisovan Hunt Down Under", *Cosmos*, 18 May 2022.

123 Ann Gibbons, "Stunning 'Dragon Man' Skull May Be an Elusive Denisovan – or a New Species of Human", *Science*, 25 June 2021; Ian Sample, "Massive Human Head in Chinese Well Forces Scientists to Rethink Evolution", *The Guardian*, 26 June 2021.

124 Dyani Lewis, "Mysterious Human Fossil Found in Taiwan Was a Denisovan", *Nature*, 10 April 2025.

125 Eleanor Scerri, "Origin of Our Species: Why Humans Were Once So Much More Diverse", *New Scientist*, 25 April 2018.

126 Michael Marshall, "DNA from Another Mystery Human Ancestor Lingers in Some People", *New Scientist*, 2 April 2018.

127 Faysal Bibi, "A Wetter Ancient Arabia Could Have Enabled Easier Intercontinental Species Dispersal", *Nature*, 9 April 2025.

128 University of Cambridge, "Genetic Study Reveals Hidden Chapter in Human Evolution", 18 March 2005, online at www.camc.ac.uk.

129 In part of greater computational power, "we're starting to see the next wave of methods development", said Joshua Akey, a professor of genomics at the Lewis-Sigler Institute for Integrative Genomics at Princeton University. "And that's allowing us to start making new inferences from the data ... that the previous generation of methods couldn't make." Jordana Cepelewicz, "Fossil DNA Reveals New Twists in Modern Human Origins", *Quanta*, 29 August 2019.

130 Pääbo, *Neanderthal Man*, p. 175.

131 Xiujie Wu and Christopher Bae, "Xujiayao *Homo*: A New Form of Large-Brained Hominin in Eastern Asia", *Paleoanthropology*, 24 March 2024, online at https://paleoanthropology.org.

132 David Reich, "How One Small Tribe Conquered the World 70,000 Years Ago", Dwarkesh Podcast, 30 August 2024, https://www.dwarkesh.com/p/david-reich.

133 T. Cousins, A. Scally and R. Durbin, "A Structured Coalescent Model Reveals Deep Ancestral Structure Shared by All Modern Humans", *Nature Genetics* 57, 2025, pp. 856–64.

134 A. Bergström et al. "Origins of Modern Human Ancestry", *Nature* 590, 2021, pp. 229–37.

135 Michael Balter, "Humans and Neandertals Likely Interbred in Middle East", *Science*, 28 January 2015.

Chapter Three

1 Iain Chalmers, "The Pre-History of the First Cochrane Centre", in F.X. Bosch (ed.), *Back to the Front* (ed.), The Cochrane Collection, Oxford, 2003, p. 243.

2 R.B. Bean and W.B. Bean, *Sir William Osler: Aphorisms from His Bedside Teachings and Writings*, Henry Schuman, 1950.

3 Martin Pengelly, "'You Are Not a Horse': FDA Tells Americans Stop Taking Dewormer for Covid", *The Guardian*, 23 August 2021. The FDA ended up having to remove its tweet after being sued. See: Stephanie Vine, "Saying Farewell to the Famous Horse Tweet", *The Medicine Maker*, 26 April 2024.

4 A review of the laetrile affair: Irving J. Lerner, "The Whys of Cancer Quackery, *Cancer* 53(s3), 1984, pp. 815–19. Another example of a thoroughly tested failure is ginkgo, the extract from the leaves of the gingko tree, which in the 1990s raised fervent hopes of staving off the dementia epidemic in ageing Western populations. "Multiple clinical trials involving thousands of patients have conclusively shown that treatment with ginkgo biloba for up to six years does not prevent cognitive decline or dementia, including Alzheimer's disease", according to https://www.alzdiscovery.org/.

5 Elizabeth Finkel, "A Letter to My Vaccine-Hesitant Friends, *Cosmos*, 14 September 2021.

6 Bill Bynum and Helen Bynum, "Object Lessons", *The Lancet* 386(9997), 5 September 2015, pp. 945–46.

7 Jason Dowling, "Demolition of Key Hotel Approved", *The Age*, 30 March 2011.

8 Lisa Schnirring, "China Releases Genetic Data on New Coronavirus, Now Deadly", *CIDRAP*, 12 January 2020; Eddie Holmes at the University of Sydney, who collaborates with Zhang, persuaded him to release the sequence sooner rather than later. That contributed to his recognition as NSW Scientist of the Year: Kate Aubusson, "Virus Rebel Professor Edward Holmes Named NSW Scientist of the Year", *The Sydney Morning Herald*, 27 October 2020.

9 Sharon Lewin, personal communication, May 2021.

10 Cate Swannell, "First Isolation of SARS-CoV-2 Outside China Wins National Research Prize", *Medical Journal of Australia*, 28 July 2021.

11 The Carter Center, River Blindness Elimination Program, online at https://www.cartercenter.org/

12 Eng Eong Ooi, "Repurposing Ivermectin as an Anti-dengue Drug", *Clinical Infectious Diseases* 72(10), 15 May 2021, pp. e594–e595.

13 Jorge Caballero quoted in Nick Robins-Early, "Desperation, Misinformation: How the Ivermectin Craze Spread across the World", *The Guardian*, 24 September 2021.

14 Harriet Alexander and Clair Weaver, "How a False Science 'Cure' Became Australia's Contribution to the Pandemic", *The Sydney Morning Herald*, 22 October 2021; Liam Mannix, "How an Australian Covid Cure Conquered the World – Despite No Robust Evidence It Works", *The Sydney Morning Herald*, 7 November 2020.

15 Aila Slisco, "Anti-Parasite Drug Used Since 1980s May Help Stop Coronavirus, New Study Says", *Newsweek*, 3 April 2020.

16 Data from the Johns Hopkins Coronavirus Resource Center (https://coronavirus.jhu.edu/data/mortality). Peru: 666 deaths per 100,000, case fatality ration (CFR) 4.9 percent; Brazil: 329 deaths per 100,000, CFR 1.9 percent; USA: 341 deaths per 100,000, CFR 1.1 percent. Australia: 77 deaths per 100,000, CFR 0.2 percent. Deaths per 100,000 represents deaths as a proportion of the entire population, tested or not. CFR indicates deaths per number of known cases. CFR could be artificially high if there is low testing, but deaths per 100,000 should provide a more accurate comparison between countries. Indeed, Peru's CFR is 2.6 times higher than Brazil's, while its cases per 100,000 is only two times higher. See also: Camila Gianella Malca et al., "How Peru Became the Country with the Highest Covid Death Rate in the World, *The Conversation*, 17 November 2021.

17 Garcia says it will be very difficult to prove. "Death certificates do not include the medication that patients were taking."

18 Zamora and Gotuzzo, quoted in Catherine Offord, "Surgisphere Sows Confusion About Another Unproven COVID-19 Drug", *The Scientist*, 16 June 2020.

19 Patricia Garcia, Zoom interview, 26 October 2021.

20 Patricia J. García, "Promoting Public Health Research in Peru", *The Lancet*, vol. 374, no. 9694, 21–27 November 2009, https://doi.org/10.1016/S0140-6736(09)61811-0.

21 Joseph Schatz, "Patricia Garcia: Promoting Public Health Research in Peru", *The Lancet* 374(9698), 17 October 2009, p. 1321.

22 P. Gautret et al., "Hydroxychloroquine and azithromycin as a treatment of Covid-19: Results on an Open-Label Non-Randomized Clinical Trial", *International Journal of Antimicrobial Agents* 56(1), July 2020 (Retracted: retraction in *International Journal of Antimicrobial Agents* 65(1), January 2025)

23 Luis Ignacio Garegnani, Eva Madrid, and Nicolás Meza, "Misleading Clinical Evidence and Systematic Reviews on Ivermectin for COVID-19", *BMJ Evidence-Based Medicine* 27(3) 2022, pp. 156–58. "An important controversial point to consider in any rationale is the 5 μM required concentration to reach the anti-SARS-CoV-2 action of ivermectin observed in vitro, which is much higher than 0.28 μM, the maximum reported plasma concentration achieved in vivo with a dose of approximately 1700 μg/kg (about nine times the FDA-approved dosification). In this sense, basic fundamentals for assessing ivermectin in COVID-19 at a clinical level appear to be insufficient. Among other reasons, we believe this might have led WHO to exclude ivermectin from its Solidarity Trial for repurposed drugs for COVID-19, which raises questions about the pertinence of conducting clinical studies on ivermectin."

24 Amit N. Patel et al., "Usefulness of Ivermectin in COVID-19 Illness", *ISGlobal*, 19 April 2020. It was originally published on the preprint server SSRN, whose credo is "Tomorrow's Research Today" and which is owned by publishing giant Elsevier and Mendeley.

25 Kari A.O. Tikkinen et al., "COVID-19 Clinical Trials: Learning from Exceptions in the Research Chaos", *Nature Medicine* 26(11), November 2020, pp. 1671–72. As noted in the article: "The speed of the RECOVERY trial was record-breaking: the period from protocol to first patient recruitment was nine days, with the 176 UK hospitals recruiting over 12,000 hospitalised patients (15 percent of UK Covid-19 cases), and it provided clear answers within a few months on the effectiveness of dexamethasone and the ineffectiveness of hydroxychloroquine and lopinavir–ritonavir."

26 Peter Horby et al., "Effect of Dexamethasone in Hospitalized Patients with COVID-19 – Preliminary Report", *medRxiv*, 22 June 2020; Peter Horby et al., "Effect of Hydroxychloroquine in Hospitalized Patients with COVID-19 – Preliminary Results from a Multi-centre, Randomized, Controlled Trial", *medRxiv*, 15 July 2020. "Patients allocated to hydroxychloroquine were less likely to be discharged from hospital alive within 28 days."

27 Martin Landray, one of the architects of the RECOVERY trial, estimates a million lives were saved as a result of the trial. See Ruth Abrahams, "A Million Reasons for Talking to Professor Martin Landray", *University of Oxford Science Blog*, 20 May 2021, online at https://www.ox.ac.uk/.

28 Carlos Javier Chaccour Diaz, Zoom interview 12 October 2021 and email 26 March 2025: "The surge in malaria cases in Venezuela in the late 2010s and early 2020s

was due to the economic crisis and people turning to mining work to improve their income. These mines are all south of the Orinoco River, most deep in the jungle in indigenous territories, the places I worked in for a few years. These mines are precarious, people live in topolin tents with no sanitation nor running water and surrounded by artificial puddles created by working the land with water cannons, in the midst of the forest this is a mosquito paradise."

29 Carlos Chaccour et al., "Effect of Ivermectin on *Anopheles gambiae* Mosquitoes Fed on Humans: The Potential of Oral Insecticides in Malaria Control", *Journal of Infectious Diseases* 202(1), 1 July 2010, pp. 113–16.

30 Carlos Chaccour et al., "Ivermectin and COVID-19: Keeping Rigor in Times of Urgency", *The American Journal of Tropical Medicine and Hygiene* 102(6), June 2020, pp. 1156–57.

31 The preprint reported on the fates of 1408 critically ill patients infected with Covid, as ascertained by a PCR test from 169 hospitals across Asia, Australia, Europe, Africa, and North and South America. All were on mechanical ventilators. Exactly half (704) were treated once with a standard dose of ivermectin. The other 704 were perfectly matched for their age, gender, race and medical conditions. Such perfect matching is unusual.

32 Melissa Davey, "Questions Raised Over Hydroxychloroquine Study Which Caused WHO to Halt Trials for Covid-19", *The Guardian*, 28 May 2020.

33 Melissa Davey and Stephanie Kirchgaessner, "Surgisphere: Mass Audit of Papers Linked to Firm Behind Hydroxychloroquine Lancet Study Scandal", *The Guardian*, 10 June 2020.

34 Chaccour quoted in Catherine Offord, "Surgisphere Sows Confusion about Another Unproven COVID-19 Drug", *The Scientist*, June 2020.

35 Chaccour quoted in Melissa Davey, "Unreliable Data: How Doubt Snowballed Over COVID-19 Drug Research That Swept the World", *The Guardian*, 4 June 2020. For more about the scientists who were involved with Surgisphere, see: Charles Piller, "Who's to Blame? These Three Scientists Are at the Heart of the Surgisphere COVID-19 Scandal", *Science*, 8 June 2020.

36 Nick Robins-Early, "Desperation, Misinformation: How the Ivermectin Craze Spread Across the World", *The Guardian*, 24 September 2021.

37 Ricardo Peña-Silva et al., "Pharmacokinetic Considerations on the Repurposing of Ivermectin for Treatment of COVID-19", *British Journal of Clinical Pharmacology* 87(3), March 2021, pp. 1589–90.

38 Carlos Chaccour, Zoom interview, 13 October 2021.

39 Patricia Garcia told me she heard this on a radio survey.

40 Mika Turkia, "A Timeline of Ivermectin-Related Events in the Covid-19 Pandemic", preprint, April 2021, doi: 10.13140/RG.2.2.13705.36966.

41 Charles Piller and Kelly Servick, "Two Elite Medical Journals Retract Coronavirus Papers Over Data Integrity Questions", *Science*, 4 June 2020. Alarmingly, despite the

retractions, many scientists were still citing these articles six months on. Charles Piller, "Many Scientists Citing Two Scandalous COVID-19 Papers Ignore Their Retractions", *Science*, 4 June 2020.

42 Catherine Offord, "Surgisphere Sows Confusion About Another Unproven COVID-19 Drug", *The Scientist*, 16 June 2020.

43 Carlos Chaccour et al., "The Effect of Early Treatment with Ivermectin on Viral Load, Symptoms and Humoral Response in Patients with Non-severe COVID-19: A Pilot, Double-blind, Placebo-controlled, Randomized Clinical Trial", *EClinicalMedicine* 32, 2021.

44 Patricia Garcia, email, 28 October 2021.

45 Nick Robins-Early, "Ivermectin Frenzy: The Advocates, Anti-Vaxxers and Telehealth Companies Driving Demand", *The Guardian*, 13 September 2021.

46 Elizabeth Finkel, "Jury's in for Ivermectin: 'No Clinical Significance'", *Cosmos Magazine*, 1 April 2022.

47 Craig Raynor, Zoom interview, 1 April 2022. For more on personal attacks on scientists, see: Cathleen O'Grady, "Overwhelmed by Hate: COVID-19 Scientists Face an Avalanche of Abuse, Survey Shows", *Science*, 24 March 2022.

48 TOGETHER was funded by the Bill and Melinda Gates Foundation, the Rainwater Foundation and FastGrants. See Reis Gilmar et al., "A Multi-Center, Adaptive, Randomized, Platform Trial to Evaluate the Effect of Repurposed Medicines in Outpatients with Early Coronavirus Disease 2019 (Covid-19) and High-Risk for Complications: the TOGETHER Mater Trial Protocol", *Gates Open Research* 5(117), 2020. Gates explicitly did not fund the ivermectin arm. Ivermectin was not included in the initial line-up of repurposed drugs because the ivermectin dose required to kill the virus was more than twenty times the maximum approved dose. But by mid 2020, ivermectin had replaced hydroxychloroquine as a popular COVID cure. Given the situation on the ground, the TOGETHER team decided to include ivermectin in their trial.

COVID-OUT, a randomised trial of metformin, ivermectin, and fluvoxamine, was funded by the Parsemus Foundation, Rainwater Charitable Foundation, Fast Grants, and UnitedHealth Group Foundation. New England Journal of Medicine, 17 August 2022.

AACTIV-6 explored up to seven repurposed drugs and was funded by the US National Institutes of Health (NIH, "Large Clinical Trial to Study Repurposed Drugs to Treat Covid-19 Symptoms", 19 April 2021, online at https://www.nih.gov/). The Oxford University–led PRINCIPLE trial – The Platform Randomised trIal of treatmeNts in the Community for epIdemic and Pandemic iLlnEsses – is funded by UK Research and Innovation (UKRI) and National Institute for Health Research (NIHR) (https://www.phctrials.ox.ac.uk/).

PRINCIPLE was designed to rapidly identify treatments that showed promise and rule out treatments that did not work. Early on in the pandemic, the antibiotics azithromycin and doxycycline were being widely used around the world to treat acute Covid-19; PRINCIPLE was the first major trial in the community to show that, in the absence of other indications, these antibiotics did not benefit Covid patients and so merely put them at unnecessary risk of side effects and added to the problem of antibiotic resistance. PRINCIPLE also found that the anti-inflammatory drug, colchicine, did not help people get better any quicker. A key finding from the trial announced in April 2021 was that the commonly used asthma drug, inhaled Budesonide, a steroid, was effective in reducing recovery time by around three days and that there was a high probability that it also reduced hospital admission.

49 Gilmar Reis et al., "Effect of Early Treatment with Fluvoxamine on Risk of Emergency Care and Hospitalisation among Patients with Covid-19: The TOGETHER Randomised, Platform Clinical Trial", *The Lancet Global Health* 10(1), January 2022, e42–e51; "Fluvoxamine Can Cut Covid-19 Hospitalisations by 30 percent", *Open Access News*, 28 October 2021; Thomas G. Stewart et al., "Higher-Dose Fluvoxamine and Time to Sustained Recovery in Outpatients With COVID-19: The ACTIV-6 Randomized Clinical Trial", *JAMA* 330(24), 26 December 2023, pp. 2354–63.

50 "Joe Rogan Brett Weinstein Pierre Kory Ivermectin Podcast Mashup", Part 2, 24 June 2021, https://www.youtube.com/watch?v=fqb6NY4Yi74.

51 Stephanie M. Lee and Ken Bensinger, "A Prominent Study Said Ivermectin Prevents COVID, but the Data Is Suspect", *BuzzFeed News*, 2 September 2021; David Gorski, "Ivermectin Is the New Hydroxychloroquine, Take 2", *Science-Based Medicine*, 21 June 2021.

52 Front Line COVID-19 Critical Care Alliance, "Dr Pierre Kory, President of the FLCCC Alliance, Testifies Before Senate Committee on Homeland Security and Governmental Affairs Looking into Early Outpatient COVID-19 Treatment", *Newswise*, 8 December 2020.

53 Kristina Fiore, "What's Behind the Ivermectin-for-COVID Buzz?", *MedPage Today*, 4 December 2020.

54 Jake Zuckerman, "Judge Orders Cinci Hospital to Treat COVID-19 Patient with Ivermectin, Despite CDC Warnings", *Ohio Capital Journal*, 30 August 2021.

55 Paul Glasziou, Zoom interviews, 21 December 2021 and 14 March 2022.

56 Andrew Hill et al., "Ivermectin for COVID-19: Addressing Potential Bias and Medical Fraud", *Open Forum Infectious Diseases* 9(2), February 2022.

57 Roger L. Sur and Philipp Dahm, "History of Evidence-Based Medicine", *Indian Journal of Urology* 27(4), October–December 2011, pp. 487–89.

58 Hilda Bastian et al., "Seventy-Five Trials and Eleven Systematic Reviews a Day: How Will We Ever Keep Up?", *PLOS Medicine* 7(9), 21 September 2010.

59 "Randomised Trial of Intravenous Streptokinase, Oral Aspirin, Both, or Neither among 17, 187 Cases of Suspected Acute Myocardial Infarction: ISIS-2", *JACC* 12, December 1988, pp. A3–A13.

60 Zygmunt Sadowski et al., "Multicenter Randomized Trial and a Systematic Overview of Lidocaine in Acute Myocardial Infarction", *American Heart Journal* 137(5), May 1999, pp. 792–98.

61 Thomas J. Moore, *Deadly Medicine: Why Tens of Thousands of Heart Patients Died in America's Worst Drug Disaster*, Simon and Schuster, 1995.

62 Benjamin Djulbegovic et al., "Trial Unpredictability Yields Predictable Therapy Gains", *Nature* 500(7463), 21 August 2013, pp. 395–96.

63 Djulbegovic et al., "Trial Unpredictability Yields Predictable Therapy Gains".

64 L. Goldman and A. Dennis, eds. *Cecil's Text Book of Medicine, Approach to Medicine, the Patient, and the Medical Profession: Medicine as Learned and Humane Profession*, Elsevier, 2004.

65 Allan Lew, telephone interview, 7 March 2022.

66 Christine McDonald, in-person interview, 24 April 2022.

67 John Matthewson, telephone interview, 19 September 2022.

68 Kristina Fiore, "What's Behind the Ivermectin-for-COVID Buzz?", *MedPage Today*, 4 December 2020.

69 A term Archie Cochrane used to describe the all-knowing attitude of some of his medical colleagues. "I suggested to them the value of controlled trials, but was horrified when they reacted like elderly physicians and headmasters. They too suffered from the God complex. They knew what to do without the help of any trials. It was depressing." Archie Cochrane with Max Blythe, *One Man's Medicine*, BMJ Books, 1989.

70 "In spite of all my troubles there I had learnt a lot about myself. I was satisfied that I had really cared for my patients as well as I could have in the circumstances. I had found it emotionally satisfying and distressing at the same time. I knew my patients so well that I was miserable when they died. I also found "caring" intellectually unsatisfactory. What I decided I could not continue doing was making decisions about intervening (for example, pneumothorax and thoracoplasty) when I had no idea whether I was doing more harm than good. I remember reading a pamphlet (I think from the BMA) extolling the advantages of the freedom of British doctors to do whatever they thought best for their patients. I found it ridiculous. I would willingly have sacrificed all my medical freedom for some hard evidence telling me when to do a pneumothorax. I feared I had shortened some lives by doing it on the wrong cases. (This was certainly the birth of an idea which culminated in *Effectiveness and Efficiency*)." Cochrane, *One Man's Medicine.*

71 P.C. Elwood et al., "A Randomised Controlled Trial of Acetyl Salicylic Acid in the Secondary Prevention of Mortality from Myocardial Infarction", *British Medical Journal* 1(5905), 9 March 1974, pp. 436–40.

72 Archie Cochrane, *Effectiveness and Efficiency: Random Reflections on Health Services*, Nuffield Provincial Hospital Trust, 1972.

73 "Furthermore, looking at the minuscule gains in life expectancy (still 77–78 for women, and 68 for men) I do think there is sufficient evidence to justify a closer look at the effectiveness of some of the therapies in use in the NHS and the efficiency with which they are applied. Before doing this, however, I want to introduce a chapter on the value of different types of evidence, because this is crucial in assessing effectiveness and efficiency. There still seems to be considerable misunderstanding amongst the general public and some medical people about the relative value of opinion, observation, and experiment in testing hypotheses." Cochrane, *Effectiveness and Efficiency.*

74 C.T. Dollery, "Constructive Attack: *Effectiveness and Efficiency*", *British Medical Journal* 56, 1972.

75 I. Chalmers, "The Prehistory of the First Cochrane Centre", in X. Bosch, ed., *Back to the Front,* The Cochrane Collection, Oxford, 2003, pp. 242–53. Referenced in Roger L. Sur and Philipp Dahm, "History of Evidence-Based Medicine", *Indian Journal of Urology* 27(4), October–December 2011, pp. 487–89.

76 Jane Harding and Caroline Crowther, "A History: Antenatal Corticosteroids", *O&G Magazine* 21(1), Autumn 2019.

77 Richard Doll and A. Bradford Hill, "Smoking and Carcinoma of the Lung", *British Medical Journal* 2(4682), 30 September 1950, pp. 739–48; Ernest L. Wynder and Evarts A. Graham, "Tobacco Smoking as a Possible Etiologic Factor in Bronchiogenic Carcinoma: A Study of Six Hundred and Eighty-Four Proved Cases", *Journal of the American Medical Association* 143(4), 27 May 1950, pp. 329–36. (The risk was twenty-five times higher in those who smoked more than twenty-five cigarettes a day than in nonsmokers.) Maria Elisa Di Cicco et al., "Mortality in Relation to Smoking: The British Doctors Study", *Breathe* 12(3), September 2016, pp. 275–76.

78 Australasian Menopause Society, "Risks and Benefits of MHT", *Australasian Menopause Society*, 2021, online at https://www.menopause.org.au/.

79 Iain Chalmers et al., "The Advent of Fair Treatment Allocation Schedules in Clinical Trials During the Nineteenth and Early Twentieth Centuries", *Journal of the Royal Society of Medicine* 105(5), May 2012.

80 James Lind, *A Treatise of the Scurvy in Three Parts*, 1753, online at The James Lind Library, jameslindlibrary.org.

81 Marcus White, "James Lind: The Man Who Helped to Cure Scurvy with Lemons", BBC News, October 2016.

82 U. Tröhler, "James Lind and Scurvy: 1747 to 1795", *JLL Bulletin: Commentaries on the History of Treatment Evaluation*, 2003, online at jameslindlibrary.org.

83 Iain Milne and Iain Chalmers, "Alexander Lesassier Hamilton's 1816 Report of a Controlled Trial of Bloodletting", *Journal of the Royal Society of Medicine* 108(2), February 2015, pp. 68–70.

84 Philippe Bautret et al., "Hydroxychlorquine and Azithromycin as a Treatment of Covid-19: Results on an Open-Label Non-Randomized Clinical Trial", *International Journal of Antimicrobial Agents* 56(1), July 2020 (retracted).

85 Julia Carrie Wong, "Hydroxychloroquine: How an Unproven Drug Became Trump's Coronavirus 'Miracle Cure'", *The Guardian*, 7 April 2020; Scott Sayare, "He Was a Science Star. Then He Promoted a Questionable Cure for Covid-19", *The New York Times Magazine*, 12 May 2020.

86 Retraction notice to "Hydroxychloroquine and Azithromycin as a Treatment of Covid-19: Results of an Open-Label Non-Randomized Clinical Trial", *International Journal of Antimicrobial Agents* 65, January 2025.

87 "Covidence explains the difference between systematic review and meta-analysis", online at Covidence.org.

88 C. Axfors et al., "Mortality Outcomes with Hydroxychloroquine and Chloroquine in Covid-19 from an International Collaborative Meta-Analysis of Randomized Trials", *Nature Communications* 12, 2021.

89 Elizabeth Finkel, "Data Detectives Dig into Ivermectin Studies", *Cosmos*, 7 October 2021; Melissa Davey, "Huge Study Supporting Ivermectin as Covid Treatment Withdrawn Over Ethical Concerns", *The Guardian*, 15 July 2021.

90 Craig Rayner, email correspondence, 5 April 2022.

91 As explained to me by Carlos Chaccour.

92 Juan Chamie, "Real-World Evidence: The Case of Peru – Causality Between Ivermectin and COVID-19 Infection Fatality Rate", 2020, online at https://www.researchgate.net/.

93 Juan J. Chamie-Quintero et al., "Sharp Reductions in COVID-19 Case Fatalities and Excess Deaths in Peru in Close Time Conjunction, State-By-State, with Ivermectin Treatments", *SSRN*, 21 January 2021.

94 Elizabeth Finkel, "What We Know About Ivermectin: Correlation Is Not Causation", *Cosmos*, September 2021.

95 Peter Walsh, Roman Biek and Leslie Real, "Wave-Like Spread of Ebola Zaire", *PLOS Biology*, October 2005.

96 Pierre Kory et al., "Review of the Emerging Evidence Demonstrating the Efficacy of Ivermectin in the Prophylaxis and Treatment of COVID-19", *American Journal of Therapeutics* 28(3), May–June 2021, pp. e299–e318.

97 Gabriela Buffet et al., "Inaccurate Real-World Data Does Not Provide Real-World Answers", *American Journal of Therapeutics* 28(5), September–October 2021, pp. e596–e598.

98 Peter Manu, "Expression of Concern for Kory P, Meduri GU, Varon J, Iglesias J, Marik PE. 'Review of the Emerging Evidence Demonstrating the Efficacy of Ivermectin in the Prophylaxis and Treatment of COVID-19'", *American Journal of Therapeutics* 29(2), March–April 2022, p. e231. Another paper with Kory as the first author was retracted. See P. Kory et al., "Clinical andScientific Rationale for the 'MATH+' Hospital Treatment Protocol for Covid-19", Journal of Intensive Care Medicine 36(2), February 2021,pp. 135–56.

99 Laura L. George et al., "Cure of Duodenal Ulcer After Eradication of Helicobacter pylori", *Medical Journal of Australia* 153(3) 6 August 1990, pp. 145–149.

100 Sabine Hazan et al., "Effectiveness of Ivermectin-Based Multidrug Therapy in Severely Hypoxic, Ambulatory Covid-19 Patients", *Future Microbiology* 17, March 2022, pp. 339–50.

101 Using a background death rate as a comparison is problematic because death rates from Covid varied according to the place and time. The UK reports the death rate for hospitalised patients in the first half of 2020 was 26 percent. Another study based on northern California's Kaiser Permanente healthcare system reported the death rate as 6.3 percent for those treated on the ward.

102 This study was funded by ProgenaBiome, LLC. S. Hazan declares that she has pecuniary interest in Topelia Pty Ltd in Australia and Topelia Pty Ltd in the USA, where development of Covid-19 preventative/treatment options are being pursued. She has also filed patents relevant to coronavirus treatments. She is the founder and owner of Microbiome Research Foundation, ProgenaBiome and Ventura Clinical Trials. TJ Borody declares that he has pecuniary interest in Topelia Pty Ltd in Australia and Topelia Therapeutics, Inc. in the USA developing Covid-19 preventative/treatment medications.

103 Kurinchi Gurasamy, Zoom interview, 11 April 2022.

104 Henry K. Beecher, "The Powerful Placebo", *Journal of the American Medical Association* 159(17), 24 December 1955, pp. 1602–6; Gunver S. Kienle and Helmut Kiene, "The Powerful Placebo Effect: Fact or Fiction?", *Journal of Clinical Epidemiology* 50(12), December 1997, pp. 1311–18.

105 "Statement of Retraction: Effectiveness of Ivermectin-Based Multidrug Therapy in Severely Hypoxic, Ambulatory Covid-19 Patients", *Future Microbiology* 9(1), January 2025.

106 Morteza Shakhsi Niaee et al., "Ivermectin as an Adjunct Treatment for Hospitalized Adult COVID-19 Patients: A Randomized Multi-Center Clinical Trial", *Asian Pacific Journal of Tropical Medicine* 14(6), June 2021, pp. 266–73.

107 Andrew Hill et al., "Ivermectin for COVID-19: Addressing Potential Bias and Medical Fraud", *Open Forum Infectious Diseases* 9(2), February 2022.

108 Melissa Davey, "Huge Study Supporting Ivermectin as Covid Treatment Withdrawn Over Ethical Concerns", *The Guardian*, 15 July 2021; Elizabeth Finkel, "Data Detectives Dig into Ivermectin Studies", *Cosmos*, 7 October 2021.

109 Sara Reardon, "Flawed Ivermectin Preprint Highlights Challenges of COVID Drug Studies", *Nature* 596(7871), 2 August 2021, pp. 173–74.

110 Gilmar Reis et al., "Effect of Early Treatment with Ivermectin among Patients with Covid-19", *New England Journal of Medicine* 386(18), 5 May 2022, pp. 1721–31; Elizabeth Finkel, "Jury's In for Ivermectin: 'No Clinical Significance'", *Cosmos*, 1 April 2022.

111 Andrew Bryant et al., "Ivermectin for Prevention and Treatment of COVID-19 Infection: A Systematic Review, Meta-Analysis, and Trial Sequential Analysis to

Inform Clinical Guidelines", *American Journal of Therapeutics* 28(4), July–August 2021, pp. e434–e460.

112 Andrew Hill et al., "Meta-Analysis of Randomized Trials of Ivermectin to Treat SARS-CoV-2 Infection", *Open Forum Infectious Diseases* 8(11), November 2021 (retracted).

113 Maria Popp et al., "Ivermectin for Preventing and Treating COVID-19", *Cochrane Database of Systematic Reviews* 7(28), July 2021.

114 Finkel, ""Data Detectives Dig into Ivermectin Studies", *Cosmos*, 7 October 2021.

115 Andrew Hill et al., "Ivermectin for COVID-19: Addressing Potential Bias and Medical Fraud", *Open Forum Infectious Diseases* 9(2), February 2022.

116 Irwin J. Lerner, "The Whys of Cancer Quackery", *Cancer* 53(3), Suppl, 1 February 1984, pp. 815–19.

117 Carolyn T. Bramante et al., "Randomized Trial of Metformin, Ivermectin, and Fluvoxamine for Covid-19", *New England Journal of Medicine* 387(7), 18 August 2022, pp. 599–610; Salim S. Abdool Karim and Nikita Devnarain, "Time to Stop Using Ineffective Covid-19 Drugs", *New England Journal of Medicine* 387(7), 18 August 2022, pp. 654–55.

118 ACTIV-6 also failed to show any effect of metformin on reducing the severity of Covid in a later analysis in January 2025: Carolyn T. Bramante et al., "Metformin on Time to Sustained Recovery in Adults with COVID-19: The ACTIV-6 Randomized Clinical Trial", *medRxiv*, 14 January 2025 (preprint).

119 Gail Hayward et al., "Ivermectin for COVID-19 in Adults in the Community (PRINCIPLE): An Open, Randomised, Controlled, Adaptive Platform Trial of Short- and Longer-Term Outcomes", *Journal of Infection* 88(4), April 2024, article 106130.

120 Ly-Mee Yu and Christopher C. Butler, "PRINCIPLE Trial, 'Inhaled Budesonide for Covid-19 in People at High Risk of Complications in the Community in the UK (PRINCIPLE): A Randomised, Controlled, Open-Label, Adaptive Platform Trial", *Lancet* 398(10303), September 2021, pp. 843–55.

121 Elizabeth Weise, "Pennsylvania Man Who Took Ivermectin to Treat COVID-19 Dies", *USA Today*, 13 December 2021.

122 Kaitlyn Tiffany, "How Ivermectin Became a Belief System", *The Atlantic*, 22 March 2023; Joel M. Levin et al., "The Political Polarization of COVID-19 Treatments Among Physicians and Laypeople in the United States", *Proceedings of the National Academy of Sciences* 120(7), 14 February 2023.

123 Zachary Smith, "The Strange Return of Ivermectin and Hydroxychloroquine: Republicans Push Drug in State Bills", *Forbes*, 3 May 2022.

124 Public Health Communications Collaborative. "RFK Jr Continues to Promote Disproven Covid-19 Cures", 20 June 2023, https://publichealthcollaborative.org/alerts/

125 Catalina Jaramillo and Kate Yandell, "RFK Jr.'s COVID-19 Deceptions", *FactCheck.org*, 11 August 2023.

126 Lauren Weber, Lena H. Sun, and David Ovalle, "10 RFK Jr. Conspiracy Theories and False Claims, in His Own Words", *The Washington Post*, 15 November 2024; Michelle Duff, "We Learned the Hard Way: Samoa Remembers a Deadly Measles Outbreak and a Visit from RFK Jr", *The Guardian*, 26 November 2024; Meredith Wadman, "Four Vaccine Myths and Where They Came From", *Science*, 27 April 2017;

127 Seth Mnookin, "How Robert F. Kennedy, Jr., Distorted Vaccine Science", *Scientific American*, 11 January 2017; Dominique Mosbergen, "What Has RFK Jr. Said About Vaccines?", *The Wall Street Journal*, 29 January 2025. Paul Offit is quoted in Michelle Duff, "'We Learned the Hard Way'".

128 Julian Nowogrodzki, "Measles Is Surging in the US: How Bad Could It Get?", *Nature* 639(7992), 19 March 2025, pp. 846–47; World Health Organization, "Measles – United States of America", *Disease Outbreak News*, 27 March 2025, https://www.who.int/; Centers for Disease Control. "Measles Cases and Outreaks", https://www.cdc.gov/.html; Texas Health and Human Services, "Texas Announces First Death in Measles Outbreak", 26 February 2025, https://www.dshs.texas.gov/

129 Liz Essley Whyte, "FDA's Top Vaccine Official Forced Out", *The Wall Street Journal*, 29 March 2025.

Chapter Four

1 Daniel Dennett, *Consciousness Explained*, Penguin Books, 1991.

2 Daniel Dennett, *Consciousness Explained*, Peguin Books, 1993, Kindle edition, p. 21.

3 James Watson, *The Double Helix*, Penguin Books, 1968, pp. 21–22.

4 Terrence Sejnowski, "Consciousness", *Daedalus* 144(1), Winter 2015, pp. 123–32.

5 J. Knight. "From DNA to Consciousness: Crick's Legacy", *Nature* 430, 2004.

6 From the epigraph of Francis Crick's *The Astonishing Hypothesis*, Simon and Schuster, 1994.

7 Crick, *The Astonishing Hypothesis*.

8 Crick, *The Astonishing Hypothesis*, epigraph to Chapter 2.

9 Francis Crick and Christof Koch, "Towards a Neurobiobological Theory of Consciousness", *Seminars in the Neurosciences* 2, pp. 263–75.

10 Sam Harris, "The Nature of Consciousness: A Conversation with Thomas Metzinger", *Making Sense*, episode 96, SamHarris.org.

11 Association for the Scientific Study of Consciousness, Twenty-Sixth Annual Meeting, 22 June 2023, New York City.

12 Hasok Chang, *Inventing Temperature*, Oxford University Press, 2004.

13 Stanislas Dehaene, *Consciousness and the Brain*, Penguin Books, 2014, p. 116.

14 D.J. Simons and C.F. Chabris, "Gorillas in Our Midst: Sustained Inattentional Blindness for Dynamic Events", *Perception* 28(9), pp. 1059–74.

15 Brian Wandell et al., "Visual Field Maps in Human Cortex", *Neuron* 56(2), pp. 366–83.

16 Gottfried Leibniz, *Monadology*, 1714.

17 "Debate on Mind–Brain Relation: Searle vs Eccles (1984)", *Philosophy Overdose*, https://www.youtube.com/watch?v=1KemhfmAsg8.

18 D. Chalmers, "Facing Up to the Problem of Consciousness", *Journal of Consciousness Studies* 2(200), 1995, pp. 200–19.

19 Per Snaprud, "The Consciousness Wager", *New Scientist* 238(3183), 23 June 2018, pp. 28–31.

20 "Can Daniel Dennett Answer the Hard Problem of Consciousness?" *Premier Unbelievable*, https://www.youtube.com/watch?v=YeNKAyp4vTQ.

21 Anil Seth, *Being You*, Faber, 2021, p. 22.

22 Daniel Dennett, "What Is Consciousness?", *Closer to Truth*, 1 July 2014, https://youtu.be/K26fo6QRc_k?si=cyLRWS4fDGZWEQkc.

23 Evidence for such humming choirs of neurons had already been reported in the visual system. See C.M. Gray et al., "Oscillatory Responses in Cat Visual Cortex Exhibit Inter-Columnar Synchronization which Reflects Global Stimulus Properties", *Nature* 338, 1989, pp. 334–337.

24 Crick and Koch, "Towards a Neurobiological Theory of Consciousness".

25 Francis Crick, "Visual Perception: Rivalry and Consciousness", *Nature* 379, 1996, pp. 485–86.

26 Knight, "From DNA to Consciousness: Crick's Legacy".

27 Personal communication, June 2023 at the ASSC.

28 Stefan Frässle et al., "Binocular Rivalry: Frontal Activity Relates to Introspection and Action but not to Perception", *Journal of Neuroscience* 34(5), 2014, pp. 1738–47.

29 As their webpage proclaims: "We fund work on subjects ranging from black holes and evolution to creativity, forgiveness, and free will ... We share Sir John Templeton's optimism about the power of the sciences and other discovery-oriented disciplines to advance our understanding of the deepest and most perplexing questions facing humankind." See www.templeton.org.

30 Although it is awarded by the Nobel Foundation, it is not one of the original five prizes bequeathed in Alfred Nobel's will. It is formally referred to as the Nobel Memorial Prize in Economic Sciences and is presented at the Nobel Prize Award Ceremony along with the others. See www.nobelprize.org.

31 Association for Psychological Science, "Adversarial Collaboration: An EDGE Lecture by Daniel Kahneman", 23 February 2023, online at https://www.psychologicalscience.org.

32 Michael Pitts, Zoom interviews, 9 May and 13 May 2023.

33 Liad Mudrik, Zoom interview, 8 March 2023.

34 Templeton World Charity Foundation, "An Adversarial Collaboration to Test Predictions of First-Order and Higher-Order Theories of Consciousness", online at www.templetonworldcharity.org.

35 Templeton World Charity Foundation, "Accelerating Research on Consciousness", online at www.templetonworldcharity.org.

36 For a detailed review, see Anil Seth and Tim Bayne, "Theories of Consciousness", *Nature Reviews Neuroscience* 23 (2022), pp. 439–52.

37 Thomas Nagel, "What Is It Like to Be a Bat?", *The Philosophical Review* 83(4), October 1974, pp. 435–50.

38 Neven, Hartmut et al., "Testing the Conjecture That Quantum Processes Create Conscious Experience", *Entropy* 26(6), 2024, p. 460.

39 Erwin Schrodinger, *"What Is Life?" and Other Scientific Essays*. Doubleday Anchor, 1944, online at https://archive.org/details/whatislifeothers00schr/. See Chapter VI: "living matter, while not eluding the 'laws of physics' as established up to date, is likely to involve 'other laws of physics' hitherto unknown, which however, once they have been revealed, will form just as integral a part of science as the former."

40 "Daniel Dennett on Consciousness, Virtual Immortality, and Panpsychism", *Closer to Truth Chats*, https://www.youtube.com/watch?v=Djqk-RexNFs.

41 Personal communication, 23 May 2023 at ASSC, New York City.

42 Hakwan Lau, personal communication, 7 January 2025.

43 Richard Brown, Hakwan Lau and Joseph E. LeDoux, "Understanding the Higher-Order Approach to Consciousness", *Trends in Cognitive Sciences* 23(9), 2019, pp. 754–68.

44 Lee D. Erman et al., "The Hearsay-II Speech-Understanding System: Integrating Knowledge to Resolve Uncertainty", *ACM Computer Surveys* 12(2), 1980, pp. 213–53.

45 From Bernard J. Baars, *A Cognitive Theory of Consciousness*, Cambridge University Press, 1988. Online at http://cogweb.ucla.edu/Abstracts/Baars_88.html.

46 René Descartes, "Traité de l'homme", in *Descartes: Oeuvres et Lettres*. Gallimard, 1937. Referenced in Stanislas Dehaene and Jean-Pierre Changeux, "Theoretical Approaches to Conscious Processing", *Neuron* 70(2), 2011, pp. 200–27.

47 H. Taine, *De L'Intelligence*. Hachette, 1870. Quoted in Stanislas Dehaene, *Consciousness and the Brain*, Penguin, 2014, p. 165.

48 Stanislas Dehaene, Michel Kerszberg and Jean-Pierre Changeux, "A Neuronal Model of a Global Workspace in Effortful Cognitive Tasks", *Procedings of the National Academy of Sciences of the United States of America* 95(24), 1998, pp. 14529–34.

49 Dehaene, *Consciousness and the Brain*, p. 137. "Why does the brain generate synchronized neuronal oscillations? … If they shout in unison … their voice is much more likely to be heard and relayed."

50 Two of the markers Dehaene identified for the consciousness workspace are highly contested. One is the 40 Hz gamma oscillation that Crick and Koch had mooted as a neural correlate of consciousness. Another is the P3B wave, a signal that appears about a third of a second (300 msecs) after a novel stimulus. Dehaene does not see that as a blow to his theory. See "Q&A", *Neuron* 112(10), 15 May 2024, pp. 1527–30.

51 John Hawks, email correspondence, February–April 2021 and Zoom interview, 17 February 2025.

52 Christof Koch, Zoom interview, 26 May 2023.

53 Neuroscientist Anil Seth, "The Worth of Wild Ideas", *Nautilus*, 27 September 2003; philosopher Tim Bayne, "On the Axiomatic Foundations of the Integrated Information Theory of Consciousness", *Neuroscience of Consciousness* 1, 2018, niy007; computer scientist Scott Aaronson, "Giulio Tononi and Me: A Phi-nal Exchange", 30 May 2014, https://scottaaronson.blog.

54 Lucia Melloni, Zoom interview, 21 June 2021.

55 According to this annecdote: "Asked in 1919 whether it was true that only three people in the world understood the theory of general relativity, [Eddington] allegedly replied: 'Who's the third?'"

56 Richard P. Feynman, "Quantum Mechanics", The Messenger Lectures, 1964, Cornell University, online at https://www.feynmanlectures.caltech.edu/.

57 Giulio Tononi, "What's the Essence of Consciousness?", *Closer to Truth*, https://www.youtube.com/watch?v=LlR9vkx6Hv8.

58 Anil Seth, *Being You*, Faber and Faber, 2021, p. 61.

59 Tononi, "What's the Essence of Consciousness?"

60 Christof Koch, Zoom interview, May 2023.

61 A. Watakabe et al., "Local and Long-Distance Organization of Prefrontal Cortex Circuits in the Marmoset Brain", *Neuron* 111(14), pp. 2258–73, 19 July 2023.

62 Christof Koch, *The Feeling of Life Itself*, MIT Press, 2019, pp. 59–63.

63 M. Massimini et al., "Breakdown of Cortical Effective Connectivity During Sleep", *Science* 309, 30 September 2005, pp. 2228–32.

64 Seth and Bayne, "Theories of Consciousness".

65 Interview with Stan Dehaene, 24–25 June 2023, ASSC, New York City.

66 Geraint Rees, "Seeing Is Not Perceiving", *Nature Neuroscience* 4, 2001, pp. 678–80.

67 See for example: Jaan Aru et al., "Distilling the Neural Correlates of Consciousness", *Neuroscience and Biobehavioral Reviews* 36(2), February 2012; Christof Koch et al., "Neural Correlates of Consciousness: Progress and Problems", *Nature Reviews Neuroscience* 17(5), 2016; Melanie Boly et al., "Are the Neural Correlates of Consciousness in the Front or in the Back of the Cerebral Cortex? Clinical and Neuroimaging Evidence", *Journal of Neuroscience* 37(40), October 2017; Brian Odegaard et al., "Should

a Few Null Findings Falsify Prefrontal Theories of Conscious Perception?", *Journal of Neuroscience* 37(40).

68 Lucia Melloni et al., "An Adversarial Collaboration Protocol for Testing Contrasting Predictions of Global Neuronal Workspace and Integrated Information Theory", *PLoS ONE* 18(2), February 2023, e0268577.

69 Rhiannon Williams, "Brain Scans Can Translate a Person's Thoughts into Words", *MIT Tech Review*, 1 May 2023.

70 "ASSC 2023: Resolving the Bet", *Event Rap*, https://www.youtube.com/watch?v=0r3s8YM5Yi4.

71 Per Snaprud, "The Consciousness Wager", *New Scientist* 238(3183), 2018, pp. 28–31.

72 Elizabeth Finkel, "'Adversarial' Search for Neural Basis of Consciousness Yields First Results", *Science*, 25 June 2023, pp. 1309–10. See also: Elizabeth Finkel and Sharon Carleton, "Theories on Consciousness Go Head-to-Head", *The Science Show*, ABC Radio National, 16 September 2023; Elizabeth Finkel, "What a Contest of Consciousness Theories Really Proved", *Quanta Magazine*, 24 August 2023.

73 Cogitate Consortium, "An Adversarial Collaboration to Critically Evaluate Theories of Consciousness", *BioRxiv*, 30 June 2023, https://doi.org/10.1101/2023.06.23.546249.

74 Cogitate Consortium, "Adversarial Testing of Global Neuronal Workspace and Integrated Information Theories of Consciousness", *Nature* 642, 2025, pp. 133–42.

75 See "Generative Adversarial Collaborations", https://sites.google.com/ccneuro.org/gac2020/.

76 Megan Peters, Zoom interview, 28 July 2023.

77 Olivia Carter, personal communication, 24 July 2023.

78 Jakob Hohwy, Zoom interview, 12 July 2023.

79 Lucia Melloni, personal communication, 27 July 2023.

80 Kahneman also prefaced that sentence by referring to the ideas of Hungarian philosopher Imre Lakatos: "Lakatos distinguishes two paths for theories that are challenged by unanticipated findings. One is progressive refinement; the other is defensive degeneration. The path on which a theory is set eventually becomes obvious to everyone – except perhaps the theorist. And this is how adversarial collaborations can advance science without adversaries changing their mind." "Adversarial Collaboration: An EDGE Lecture by Daniel Kahneman", online at www.edge.org.

81 Christof Koch, Zoom interview, 26 May 2023.

82 Michael Strevens, *The Knowledge Machine*. Allen Lane, 2020, pp. 68–73.

83 I.I.T-Concerned, S.M. Fleming et al., "The Integrated Information Theory of Consciousness as Pseudoscience", *PsyArXiv*, posted 16 September 2023, https://doi.org/10.31234/osf.io/zsr78. Ripostes: Tim Bayne, "Nobody Knows How Consciousness Works – but Top Researchers Are Fighting Over which Theories Are Really Science",

The Conversation, 27 September 2023. A. Gomez-Marin, "The Consciousness of Neuroscience", *eNeuro* 10(11), 14 November 2023. A.K. Seth, "The Worth of Wild Ideas", *Nautilus*, 27 September 2023.

84 IIT-Concerned, M. Klincewicz et al., What Makes a Theory of Consciousness Unscientific? *Nature Neuroscience* 28, 2025, pp. 689–93.

85 "the entire cosmos is suffused with sentience": Christof Koch penned this phrase in his book *Consciousness: Confessions of a Romatic Reductionist.* MIT Press, 2012, p. 132. See also a lively debate on IIT here: John R. Searle, "Can Information Theory Explain Consciousness", *The New York Review*, 10 January 2013; and reply by Christof Koch and Giulio Tononi, "Can a Photodiode Be Conscious?", *The New York Review*, 7 March 2013.

86 A. Gomez-Marin and A.K. Seth, "A Science of Consciousness Beyond Pseudo-Science and Pseudo-Consciousness", *Nature Neuroscience* 28, 2025, pp. 703–6.

87 Hans Israel et al., *One Hundred Authors Against Einstein*, 1931, English translation at: https://archive.org/details/einstein-1.

Chapter Five

1 From Lincoln Barnett, *The Universe and Dr Einstein* (1950), quoted in Susan Ratcliffe (ed.), *Oxford Essential Quotations*, Fourth Edition, Oxford University Press, 2016.

2 https://www.intelligencetest.com/quotes/.

3 Stephen DiNardo et al., "Two-tiered Regulation of Spatially Patterned Engrailed Gene Expression During Drosophila Embryogenesis", *Nature* 332, 1988, pp. 604–9.

4 "I want to know how God created this world. I'm not interested in this or that phenomenon, in the spectrum of this or that element. I want to know His thoughts; the rest are details." From Esther Salaman, "A Talk with Einstein", *The Listener* 54, 1955, pp. 370–71, quoted in Max Jammer, *Einstein and Religion*, Princeton University Press, 1999, p. 123.

5 John Wheeler quoted in James Gleick, *Genius: The Life and Science of Richard Feynman*, Vintage Books, 1993, p. 93.

6 "probably the greatest scientific discovery ever made" is from Paul Dirac; "the greatest feat of human thinking about nature" is from Max Born; both quoted in Walter Isaacson, *Einstein: His Life and Universe*. Simon and Schuster, 2007, p. 223. "One of the greatest achievements of human thought" is from J.J. Thomson, quoted by Joel Achenbach, "100 Years Ago, an Eclipse Changed the Known Laws of Physics and Made Einstein Einstein", *The Washington Post*, 28 May 2019. To these quotes from Nobel Prize–winners, I'd add this one from physicist and populariser Brian Greene, who made a PBS documentary to celebrate the one-hundredth birthday of Eddington's confirmation of General Relativity: "yet it is in that singular achievement that we recognise the depth of the human drive for coherence, for unity". Brian Greene, *Light Falls: Space, Time and an Obsession of Einstein*, online at www.pbs.org.

7 Robyn Arianrhod, *Einstein's Heroes,* University of Queensland Press, 2003, p. 117.

8 James Clerk Maxwell, "A Dynamic Theory of the Electromagnetic Field", *Philosophical Transactions* 155, 31 December 1865, pp. 459–512.

9 Philip Gibbs, "How Is the Speed of Light Measured?", 1997, online at https://math.ucr.edu/.

10 Isaacson, *Einstein: His Life and Universe*, p. 292.

11 As Newton wrote in *Principia* in 1687, "Absolute space, in its own nature, without regard to anything external, remains always similar and immovable."

12 Bernard Schutz, personal communication, 30 January 2022: "Einstein's pre-WWI light-bending prediction was only half of the one he got when he finally got the right field equations in 1915. Had the eclipse expeditions succeeded in 1914, he would have been proved wrong!"

13 Abraham Pais, *Subtle Is the Lord: The Science and Life of Albert Einstein,* Oxford University Press, 1982, p. 30.

14 Daniel Kennefick, "Einstein Versus the Physical Review", *Physics Today* 58(9), 2005, pp. 43–48.

15 Kennefick, "Einstein Versus the Physical Review"; Jorge L. Cervantes-Cota et al., "A Brief History of Gravitational Waves", *Universe* 2(3), 2016, p. 22; Paul M. Sutter, "Could Gravitational-Wave 'Memories' Prove Einstein Wrong?", *Scientific American,* 8 April 2024.

16 Letter from K. Schwarzschild to Einstein, 22 December 1915, in "The Collected Papers of Albert Einstein", vol. 8a, doc. 169, as cited in "Schwarzschild, Master of the Universe", *Infinity Plus One* (blog), 6 May 2017, online at infinityplusonemath.wordpress.com.

17 Jeremy Bernstein, "The Reluctant Father of Black Holes", *Scientific American*, 1 April 2007; Ben Platts-Mills, "Einstein and His Peers Were 'Irrationally Resistant' to Black Holes. This Illustrated Story Explores Why", BBC, 27 June 2024.

18 Graham Phillips and Elizabeth Finkel, "Probing Creation with Waves", *Cosmos,* 2 June 2016.

19 There are doubts as to the authenticity of this advertisement. See Colin Schultz, "Ernest Schackleton's Job Ad ... Is Probably a Myth", *Smithsonian Magazine,* 10 September 2013.

20 Kip Thorne, *Black Holes and Time Warps*, Norton, 1994, p. 199.

21 Bernard Schutz, personal communication, January 2022.

22 "White Dwarf Stars", *Nature* 129, 1932, p. 803.

23 A.S. Eddington, *Stars and Atoms*, Clarendon Press, 1927.

24 "Profile: Fritz Zwicky's Extraordinary Vision", from Steven Soter and Neil deGrasse Tyson (eds), *Cosmic Horizons: Astronomy at the Cutting Edge*, New Press, 2000. Online at www.amnh.org.

25 Thorne, *Black Holes and Time Warps*, p. 164.

26 L. Landau, "Origin of Stellar Energy", *Nature* 141, 1938, pp. 333–4; Thorne, *Black Holes and Time Warps*, pp. 178–87.

27 Thorne, *Black Holes and Time Warps*, p. 193.

28 Goethe-Universität Frankfurt am Main. "How Massive Can Neutron Stars Be?", *ScienceDaily*, 16 January 2018.

29 Thorne, *Black Holes and Time Warps*, p. 217.

30 J.R. Oppenheimer and H. Snyder, "On Continued Gravitational Contraction", *Physical Review* 56, 1 September 1939, p. 455.

31 NASA, "What Are Black Holes?", Universe Forum, Harvard Smithsonian Center for Astrophysics, 1999. Online at https://lweb.cfa.harvard.edu/

32 "The Einstein Letter That Started It All", *The New York Times*, 2 August 1962, p. 169.

33 Fifteen scientists (eleven physicists and four chemists) who had won or would win the Nobel Prize emigrated from Germany shortly after Hitler came to power, most of them in 1933. These fifteen scientists were: Hans Bethe, Felix Bloch, Max Born, Albert Einstein, James Franck, Heinrich Gerhard Kuhn, Peter Debye, Dennis Gabor, Fritz Haber, Gerhard Herzberg, Victor Hess, George de Hevesy, Erwin Schrödinger, Otto Stern and Eugene Wigner. Of twenty-six German nuclear physicists cited in the literature before 1933, 50 percent emigrated.

34 McGeorge Bundy, "Hitler and the Bomb", *The New York Times Magazine*, 13 November 1988, section 6, page 42.

35 Personal communication, Bernard Schutz, January 2022.

36 Obituary, "John Archibald Wheeler, 1911–2008", Princeton Physics, online at https://phy.princeton.edu/.

37 Thorne, *Black Holes and Time Warps*, p. 223.

38 "In some sort of crude sense which no vulgarity, no humour, no overstatement can quite extinguish, the physicists have known sin; and this is a knowledge which they cannot lose." J. Robert Oppenheimer, lecture at Massachusetts Institute of Technology, 25 November 1947. In Susan Ratcliffe (ed.), *Oxford Essential Quotations*, fourth edition, Oxford University Press, 2016.

39 Lawrence S. Wittner, "Einstein's Postwar Campaign to Save the World from Nuclear Destruction", *Foreign Policy in Focus*, 1 March 2024.

40 Wittner, "Einstein's Postwar Campaign to Save the World from Nuclear Destruction"; and "Remembering Szilard: 75th Anniversary of Einstein-Szilard Letter to Roosevelt", Center for Arms Control and Non-Proliferation, 4 August 2014. Online at armscontrolcenter.org.

41 Daryl Kimball, "Oppenheimer, the Bomb, and Arms control, Then and Now", *Bulletin of the Atomic Scientists*, 29 July 2023.

42 Thorne, *Black Holes and Time Warps*, p. 226.

43 Howard Morland, "The Holocaust Bomb: A Question of Time", 1999. Online at https://fas.org/sgp/eprint/morland.html.

44 Ethan Siegel, "How Do the Most Massive Stars Die: Supernova, Hypernova, or Direct Collapse?", *Forbes*, 4 May 2018; Thorne, *Black Holes and Time Warps*, p. 197.

45 Thorne, *Black Holes and Time Warps*, p. 238.

46 Thorne, *Black Holes and Time Warps*, pp. 199–205.

47 "John Archibald Wheeler, 1911–2008", Princeton Physics. Both quotes are from Wheeler's autobiography *Geons, Black Holes and Quantum Foam: A Life in Physics*, written with his former student, physicist Kenneth Ford. Norton, 1998.

48 "Einstein thought gravitational waves were too weak to detect, and didn't believe in black holes. But I don't think he'd have minded being wrong!" Bruce Allen, managing director of the Max Planck Institute for Gravitational Physics, in "Gravitational Waves Detected 100 Years After Einstein's Prediction", LIGO, 11 February 2016, online at https://www.ligo.caltech.edu/.

49 Jorge L. Cervantes-Cota et. al, "A Brief History of Gravitational Waves", *Universe* 2(3), 2016. The crucial contribution to emerge from this conference was Feynman's thought experiment, which triggered Herman Bondi's math. The net result was the conclusion that gravitational waves carried energy and would be detectable.

50 Richard Feyman, quoted in Marco Di Mauro et al., "Towards Detecting Gravitational Waves: A Contribution by Richard Feynman", Cornell University, 31 October 2021, arXiv:2111.00330.

51 Rory Smith, personal communication, October 2022.

52 Thorne, *Black Holes and Time Warps*, p. 288.

53 A. Hewish, S. Bell et al., "Observation of a Rapidly Pulsating Radio Source", *Nature* 217, 1968, pp. 709–13.

54 US Department of Energy, "DOE Explains ... Neutron Stars", online at https://www.energy.gov/.

55 Louise Walsh, "Journeys of Discovery: Jocelyn Bell Purnell and Pulsars", online at https://www.cam.ac.uk/.

56 P.C. Peters and J. Mathews, "Gravitational Radiation from Point Masses in a Keplerian Orbit", *Physical Review* 131(435), July 1963.

57 Thorne, *Black Holes and Time Warps*, p. 288.

58 Thorne, *Black Holes and Time Warps*, p. 369.

59 Thorne, *Black Holes and Time Warps*, p. 389.

60 B. Schutz, "Determining the Hubble Constant from Gravitational Wave Observations", *Nature* 323, 1986, pp. 310–11.

61 That our universe arose from a singularity was proven mathematically by Steve Hawking. Natalie Wolchover, "Physicists Debate Hawking's Idea That the Universe Had No Beginning", *Quanta*, June, 2019.

62 Drever, the natural third contender, passed away at the age of eighty-five in March in 2017, seven months before the awarding of the prize.

63 To speed things up, Thorne made a bet that the experimenters would beat the theorists. "I fervently hoped to lose, since the simulations would be crucial to extracting the information carried by the observed waves. By early 2002, the Alliance [of theorists] had made much progress, but was still unable to simulate a full orbit of two black holes around each other. The computer codes would crash before an orbit was complete, and I was worried I might win the bet." Kip S. Thorne, Nobel Prize Lecture 2017, online at https://www.nobelprize.org.

64 Thorne, Nobel Prize Lecture 2017.

65 "The Reasons We Work in Numerical Relativity", Simulating Extreme Spacetimes (SXS), 31 January 2013, online at https://www.black-holes.org/.

66 Bernard Schutz, personal communication, January 2022.

67 Adrian Cho, "Here's the First Person to Stop Those Gravitational Waves", *Science*, 11 February 2016.

68 Graham Phillips and Elizabeth Finkel, "Probing Creation with Waves", *Cosmos*, 2 June 2016.

69 Adrian Cho, "Curtain Falls on Controversial Big Bang Result", *Science*, 30 January 2015.

70 Thorne, *Black Holes and Time Warps*, foreword.

71 LIGO, https://www.ligo.caltech.edu/page/faq.

72 Adrian Cho, "Merging Neutron Stars Generate Gravitational Waves and a Celestial Light Show", *Science*, 16 October 2017.

73 Dennis Overbye, "LIGO Detects Fierce Collision of Neutron Stars for the First Time", *The New York Times*, 16 October 2017.

74 Ethan Siegel, "Light and Gravitational Waves Don't Arrive Simultaneously", *Big Think*, 26 October 2023.

75 The LIGO Scientific Collaboration et al., "A Gravitational-Wave Standard Siren Measurement of the Hubble Constant", *Nature* 551, 2017, pp. 85–88.

76 Susan Scott, personal communication, 3 August 2020.

Epilogue

1 Alexandra Witze, Dan Garisto and Jeff Tollefson, "Five Key Climate and Space Projects on Trump's Chopping Block", *Nature*, 14 April 2025; "Trump's Proposed Budget Would Mean 'Disastrous' Cuts to Science", Science, 2 May 2025.

2 Simar S. Bajaj, "Trump Blew Up the Global Fight Against AIDS. Can It Recover?" *Wired*, 30 April 2025; Theodore Schleifer and Stephanie Nolen, "Gates Foundation Is Rattled by Trump's Threat to Its Mission", *The New York Times*, 30 April 2025.

3 Trump's proposed budget would mean 'disastrous' cuts to science, *Science*, 2 May 2025.

4 Ekaterina Galkina Cleary et al., "Comparison of Research Spending on New Drug Approvals by the National Institutes of Health vs the Pharmaceutical Industry, 2010–19", *JAMA Health Forum* 4(4), April 2023.

5 Francis Collins et al., "The NIH-Led Research Response to Covid-19", *Science* 379(6631), 2 February 2023, pp. 441–44.

6 Dell Cameron and Emily Mullin, "Doctor Behind Award-Winning Parkinson's Research Among Scientists Purged from NIH", *Wired*, 1 April 2025; Sara Reardon, "After 'Coding Error' Triggers Firings, Top NIH Scientists Called Back to Work", *Science*, 3 April 2025.

7 Jeffrey Mervis, "Exclusive: NSF Director to Resign Amid Grant Terminations, Job Cuts, and Controversy", *Science*, 24 April 2025.

8 David Frum, "America's Pro-Disease Movement", *The Atlantic*, 30 April 2025.

9 Fareed Zakaria, "Trump Is Destroying 100 Years of Competitive Advantage in 100 Days", *The Washington Post*, 25 April 2025.

10 Jeff Tollefson, Dan Garisto and Heidi Ledford, "Will US Science Survive Trump 2.0?", *Nature*, 29 April 2025.

11 Jon Cohen, "Saying 'Pandemic Is Over,' NIH Starts Cutting Covid-19 Research", *Science*, 25 March 2025.

12 Tom Bartlett, "What RFK Jr Told Grieving Texas Families About the Measles Vaccine", *The Atlantic*, 10 April 2025.

13 Rob Stein, "The Top FDA Vaccine Official Is Forced Out, Cites RFK Jr's 'Misinformation and Lies'", NPR, 28 March 2025; Katherine Dillinger, "Dramatic HHS Cuts Leave US 'Weaker as a Nation,' Says Top FDA Vaccine Official Forced Out Under Kennedy", CNN, 5 April 2025.

14 Rachel Maddow, "Donald Trump Fired the Man Who Likely Saved His Life", MSNBC, 3 April 2025.

15 Michelle Duff, "'We Learned the Hard Way': Samoa Remembers a Deadly Measles Outbreak and a Visit from RFK Jr", *The Guardian*, 26 November 2024.

16 Centers for Disease Control, "Measles Cases and Outbreaks", online at https://www.cdc.gov/.

17 Sven Bölte, "Why Do False Claims That Vaccines Cause Autism Refuse to Die? Here Are Nine Reasons", *The Conversation*, 15 January 2025.

18 David Geier, "Kennedy Turns to a Discredited Vaccine Skeptic for Autism Study", *The New York Times*, 27 March 2025; Lena H. Sun and Fenit Nirappil, "Vaccine Skeptic Hired to Head Federal Study of Immunizations and Autism", *The Washington Post*, 25 March 2025.

19 Sumathi Reddy, "If Vaccines Don't Cause Autism, What Does?", *The Australian*, 2 April 2025.

20 Brenda Goodman and Neha Mukherjee, "Kennedy Touts Unconventional Therapies for Measles Linked to Disciplined Texas Doctor", CNN, 6 March 2025.

21 Tim Henderson, "As Measles Outbreaks Grow, Trump Cuts Hinder Vaccination Efforts", *Stateline*, 2 May 2025.

22 Kevin Breuninger and Dan Mangan, "Federal Authorities Arrest Two Judges, in Escalation of Trump Immigration Crackdown", CNBC, 25 April 2025.

23 Patrick Smith and Gary Grumbach, "Trump Signs Executive Order to Stop Federal Funding for NPR and PBS", NBC News, 2 May 2025; Adam Gabbatt, "'Revenge Is His Number One Motivation': How Trump Is Waging War on the Media", *The Guardian*, 30 March 2025.

24 Courtney Kube and Dan De Luce, "National Security Agency Chief and Deputy Director Dismissed", NBC News, 4 April 2025.

25 John S. Gardner, "The Constitution of Knowledge Review: Defending Truth from Trump", *The Guardian*, 26 June 2021.

26 Kai Kupferschmidt, "Trump's Team, Often Accused of Spreading Misinformation, Slashes Misinformation Research", *Science*, 30 April 2025.

27 Mal Warwick, "The Scientists Hitler Drove Away Helped the Allies Win", Medium.com, 13 June 2024.

28 H. Holden Thorp, "Convergence and Consensus", *Science* 388(6745), 24 Apr 2025.

29 Andrew Bryant, Theresa A. Lawrie and Edmund J. Fordham, "Ivermectin for Prevention and Treatment of COVID-19 Infection: A Systematic Review, Meta-analysis, and Trial Sequential Analysis to Inform Clinical Guidelines", *American Journal of Therapeutics* 28, July 2021, e434–e460; Monica Hesse, "Scroll Through Trump's New Covid Website – and Have Your Mind Blown", *The Washington Pos*t, 23 April 2025.

30 Elizabeth Finkel, "A Letter to My Vaccine-Hesitant Friends", *Cosmos*, 14 September 2021; Elizabeth Finkel, "Adverse Reactions: Guillain Barre, TTS and the Fine Mesh Net", *Cosmos*, 14 September 2021; Elizabeth Finkel, "The Rollout: The Danger of Vaccines vs the Danger of COVID", *Cosmos*, 14 September 2021.

31 "Lab Leak: The True Origins of Covid-19", https://www.whitehouse.gov/lab-leak-true-origins-of-covid-19/; Hesse, "Scroll Through Trump's New Covid Website – and Have Your Mind Blown".

Index

Page numbers in **bold** refer to images and captions.

www.ingramcontent.com/pod-product-compliance
Lightning Source LLC
LaVergne TN
LVHW042005080726
842905LV00028B/256
9781760642716